EARLY WORK
OF
WILLARD GIBBS

PUBLICATION NO. 17
HISTORICAL LIBRARY
YALE UNIVERSITY SCHOOL OF MEDICINE

ISSUED IN CONNECTION WITH THE
CENTENNIAL ANNIVERSARY
OF THE
SHEFFIELD SCIENTIFIC SCHOOL
YALE UNIVERSITY
1847–1947

JOSIAH WILLARD GIBBS

From a daguerreotype taken during his student days

Courtesy of Miss Theodora Van Name

THE EARLY WORK OF Willard Gibbs IN APPLIED MECHANICS

COMPRISING THE TEXT
OF HIS HITHERTO UNPUBLISHED
PH.D. THESIS AND ACCOUNTS OF
HIS MECHANICAL INVENTIONS

ASSEMBLED BY
Lynde Phelps Wheeler
Everett Oyler Waters
&
Samuel William Dudley

HENRY SCHUMAN
NEW YORK

Printed in the United States of America
By George Banta Publishing Company, Menasha, Wisconsin

PREFACE

JOSIAH WILLARD GIBBS, Professor of Mathematical Physics at Yale University from 1871 until his death in 1903, is undoubtedly the most outstanding scientist this country has yet produced. His manifold contributions revealing rare analytical and creative ability, particularly in the field of physical chemistry, caused his contemporaries both here and abroad to rank him with Isaac Newton and James Clerk Maxwell. Born in 1839, the only son of the erudite Professor of Sacred Literature (of the same name), Gibbs grew up amongst academic surroundings in New Haven. Except for a few years of study in Europe he spent his whole life at Yale. The forces which left their impress upon him during his formative years are o particular interest to anyone who would trace his intellectual development to its full stature.

Gibbs was graduated from college in 1858, eleven years after the first graduate scientific school in the country had been established at Yale. This school had grown out of a small group of advanced students who had gathered about Benjamin Silliman, the Professor of Chemistry and Geology. With the backing of President Dwight, Silliman had begun teaching science at Yale in 1804 and during forty years of academic work, along with experience as consultant to mining companies and industrial concerns, he had come to appreciate the importance of the practical applications of science then so acutely needed for the development of our natural resources.

Young Gibbs had enrolled in the school after his graduation and the emphasis upon scientific applications led him to concentrate his energies upon immediately practical problems such as the form of teeth in spur gearing, brakes for railway cars, and automatic governors for steam engines. It was his thesis on the first of these subjects which brought him his Ph.D. in 1863—

the fifth such degree to be granted in the United States. That these early studies in applied mechanics should have led Willard Gibbs into the realm of higher mathematics and the more advanced phases of physical chemistry, which have so notably influenced thought both in physics and physiology, is a circumstance of challenging interest to the historian of science. To Dr. Lynde Wheeler, who conceived the plan of bringing out at the time of the Sheffield Centenary celebrations this hitherto unpublished thesis as part of a trilogy on applied mechanics, and to Professor Waters and Dean Dudley for their illuminating contributions, warm thanks are extended.

The Historical Library has considered it a privilege to sponsor this important publication because of its intimate Yale associations and its significance as a landmark in the historical development of the sciences in this country. I must thank Professor Sinnott, Director of the Sheffield Scientific School, and the members of the Sheffield Centenary Committee for financial assistance, Professor Ralph G. Van Name, nephew of Willard Gibbs, for a generous contribution toward the costs of publication, and his sister, Miss Theodora Van Name, for making available the daguerreotype that has been used as the frontispiece of the volume. The book has been designed by Carl Purington Rollins, Printer to Yale University. We are grateful to him and to the George Banta Publishing Company for encompassing the difficult details of production in a brief time.

J. F. F.

Historical Library,
Yale University,
June, 1947.

CONTENTS

N.B.: The Contributing Editors received their training in the Sheffield Scientific School, and the degrees listed from Yale University.

INTRODUCTION

LYNDE PHELPS WHEELER

It may surprise those who know Gibbs solely through his monumental work on theoretical physics to learn that during the five years of his graduate studies at Yale he was listed as a student of *engineering* in the department that soon thereafter was named the Sheffield Scientific School, or indeed that he was ever interested in the problems of applied physics. Yet if one considers the environment and the atmosphere of the times in the United States in the fifties and sixties of the last century it is not difficult to understand how anyone with a natural aptitude for science would most naturally approach its study via its applications.

The great engineering achievements of the age of rapid expansion of transportation facilities, the growth in the economic importance of machine-made goods, and the dawning realization of the aid that chemistry might furnish to agriculture and industry were all tending to focus public attention on the engineer and the possibilities of applied science. This awakening to the importance of science in the economic life of the nation had been reflected in our colleges by the first radical changes in their curricula which had taken place since their establishment in colonial days. At Yale this revolutionary academic development, fostered by nearly half a century of patient and effective missionary work by the Sillimans, culminated in the founding of the department whose centennial is being celebrated in this year 1947. In 1858, Mr. Joseph Earl Sheffield, a pioneer canal and railroad builder of New Haven, Connecticut, provided the struggling Yale Scientific School with a building and much-needed equipment. The School had been created four years before by the combination of the Yale School of Applied

Chemistry, "embraced within the Department of Philosophy and the Arts," established in 1847, and the School of Engineering, organized in 1852. In 1861 Mr. Sheffield supplemented his previous gifts by a substantial endowment, and in recognition of his generous support and encouragement the name of the School was changed to the Sheffield Scientific School.

At the time of Gibbs' graduation from college (1858) all of the available opportunities for advanced work in science, with the exception of those in mathematics and astronomy, were confined to the new department. The character of the courses offered in the new school, while loosely organized from a modern standpoint and although permitting a considerable latitude of choice on the part of the student, were at the same time of a predominantly "practical" nature in conformity with the spirit of the times.

Under these circumstances therefore it should not surprise us to find that a student of Gibbs' outstanding ability should have produced a doctor's dissertation entitled "On the Form of the Teeth of Wheels in Spur Gearing," or that he should have invented a railway car brake or an improved centrifugal governor for steam engines.

However, the contrast between this early interest in applied mechanics and his later preoccupation with the more recondite and fundamental branches of physics does present a challenging problem to one interested in the study of the development of genius. What influences turned the promising young inventor of the sixties into the profound physical philosopher of the seventies? Are the mental qualities requisite for the later development traceable in the earlier work?

An obvious explanation of this contrast is furnished by the influence of Gibbs' student days in France and Germany. In the European lecture rooms he pursued courses predominantly theoretical in nature and encountered an atmosphere of interest in science for its own sake which at that time was almost nonexistent in his own country. But while the new horizons which must have been opened up to Gibbs by his experiences abroad may have been responsible to a considerable degree for the shift of his interest from the practical to the theoretical, this fact alone does not seem to me to furnish a complete answer to the problem.

A recent re-reading of all of his published work, together with a knowledge of him gained in his classroom and study and with my recollections refreshed and made more vivid by a perusal of my own old lecture notes taken in the nineties, have convinced me that there is really no conflict between his youthful interests as revealed by the early work here first presented and his later famous contributions. It seems to me that Gibbs' abiding intellectual interest traceable through all of his work from first to last lay in the methods and problems of *mechanics*. The conditions of energy equilibrium formed the basis of his brilliant generalization of the thermodynamics of simple systems to those of complex systems. Similarly he utilized the mechanical principle of the equilibrium between the maximum kinetic and potential energies in a vibrating system to deduce the optical properties of matter on the electromagnetic theory with a minimum of assumptions as to the structure of either matter or electricity. His approach to the idea of dielectric displacement was distinctly mechanical. And here in this early work we find him generalizing the principle of torque equilibrium involved in the conical pendulum to eliminate the main disability of the device in its practical application. Throughout his whole work the same mental bent and the same methods of attack are discernible.

It is with the idea of more completely documenting this essential continuity of Gibbs' thought that it has been considered worthwhile to put these early efforts of the youthful student into the record. While neither as contributions to the sum of knowledge nor as adding materially to his reputation can these three previously unpublished items of Gibbsiana be considered to be of great importance, nevertheless the quality of the mind revealed in them is indubitably impressive and one with that of the mature Gibbs. In addition they go far to explain how it came about that, without having published a single article, Gibbs' quality had so impressed itself on the Yale faculty of the time that he was named as the first incumbent of the first chair of mathematical physics to be established in this country.

A further brief word of explanation with regard to each of these three contributions is necessary.

The thesis, for which Professor Waters has written an expository commentary, is in Gibbs' own handwriting, easily recognizable as the youthful version of that shown in the later manuscripts and letters which have survived. The title page is, however, missing. For this reason some question arose after his death as to whether this actually was the dissertation. To settle the matter as far as was possible under the circumstances, two of Gibbs' colleagues on the mathematical faculty—the late Professors William Beebe and Andrew W. Phillips—gave the manuscript a critical examination and have appended to the document a statement that "the evidence justifies us in taking this paper as Professor Gibbs' thesis for his degree as Doctor of Philosophy."

With regard to the railway brake invention it should be stated that notes concerning it and other mechanical devices, as well as correspondence in connection with the prosecution of the patent, have survived and are preserved in the Yale University Library. The original drawings for the specification are also in existence and are here reproduced. Dean Samuel W. Dudley of the School of Engineering has been over this material and has prepared a description of the invention and discussed its significance both in the history of the art and in the development of the inventor's genius.

As to the conical pendulum governor, the only information concerning it in print is contained in a single sentence in the "Biographical Memoir of Josiah Willard Gibbs" by Charles S. Hastings,* which I quote verbatim. "He is known to have made

* Nat. Acad. Sci., Biog. Mems., Vol. 6, pp. 375–393, 1909.

two interesting inventions in applied mechanics—one a brake for railway cars, which was patented (U.S. patent, No. 53971, April 17, 1866) and another, a type of governor of a higher order of approximation to astaticism than any of its predecessors, which was constructed in the machine shop of the Sheffield Scientific School and constitutes a valued apparatus in the collection of the Department of Physics."

At the time he was writing this memoir Professor Hastings showed me the model, but I do not remember to have seen it after the consolidation of the College and Scientific School Departments of Physics in the new Sloane Laboratory (1912) until some two years ago when, becoming interested in collecting biographical material about Gibbs, I suggested that a search for it be made. This was successful, and the model is now permanently on exhibition in the Sloane Laboratory and temporarily during the centennial celebration in the Yale University Library. A photograph of it is shown here facing page 64.

It is unfortunate that owing to the absence of any written record, except that quoted above, this invention cannot be precisely dated. However from evidence in the Sheffield Scientific School catalogues of the period it appears that there were no machine shop facilities available there until 1872. This fact, together with the date (1874) of the appearance of the first of the series of great thermodynamic studies, suggests that the invention is probably to be assigned to the early seventies. For if it had occurred much before Gibbs became absorbed in his great creative work in thermodynamics, it would be natural to expect that he would have taken steps to secure a patent as he had done in the case of the car brake, while if it had occurred later, he might have used the idea involved in the governor as a lecture room example of the method employed, but it is difficult to believe that he would have gone so far as to have had the model constructed.

The analysis of the Gibbs governor here presented is possibly more elaborate than the intrinsic importance or the contemporary interest of the subject justifies. I have been influenced in its preparation mainly by two considerations. The first is the novelty and originality of Gibbs' attack on a problem which had engaged the best efforts of engineers for about a century, and second, the relation of that method of attack to those he used later in his great essays in theoretical physics. I have striven to be as concise as possible consistent with an adequate documentation of an early work of genius which has survived only as a mechanical model.

February, 1947.

ON THE FORM OF THE TEETH OF WHEELS IN SPUR GEARING

By

WILLARD GIBBS

1.

On the Form of the Teeth of Wheels in Spur Gearing.

The investigation of the necessary relations between the surfaces of the teeth of wheels, in order that they may satisfy the various conditions of practical utility, is one of considerable practical importance, and is suggestive of numerous interesting geometrical or kinematical propositions. I shall confine myself to the discussion of a single case of the general subject,—that of Spur Gearing, in which the axes of the wheels are parallel, and the acting faces of the teeth are such as may be described by a plane section of the same, perpendicular to the axes moving ~~perpendicular to the axes~~ without rotation in the direction of the axes. In this case, any such section may be taken to represent the whole, and the subject reduces to one of plane geometry, in which point of view I shall consider it.

FIRST PAGE OF THE THESIS

In the author's handwriting

ON THE FORM OF THE TEETH OF WHEELS IN SPUR GEARING

§1. The investigation of the necessary relations between the surfaces of the teeth of wheels, in order that they may satisfy the various conditions of practical utility, is one of considerable practical importance, and is suggestive of numerous interesting geometrical or kinematical propositions. I shall confine myself to the discussion of a single case of the general subject, that of Spur Gearing, in which the axes of the wheels are parallel, and the acting faces of the teeth are such as may be described by a plane section of the same perpendicular to the axes moving without rotation in the direction of the axes. In this case, any such section may be taken to represent the whole, and the subject reduces to one of plane geometry, in which point of view I shall consider it.

PITCH CURVES

§2. Let two wheels, M, N, revolve about their axes m, n. Imagine an indefinite plane, coinciding with the plane of rotation, attached to each wheel and revolving with it. These planes we may also call, M, N, as they may be considered as forming part of the wheels, but without the material attribute of impenetrability. Let μ, ν be the varying angles made by a radial line in each plane with the line of centers, mn. In regard to the signs of these angles, take either direction of revolution arbitrarily as positive for μ, and the reverse as positive for ν, as in Fig. 1. If we divide mn into segments mo, no, inversely proportional to the angular velocities $\dot{\mu}$, $\dot{\nu}$, the points of M and N passing through o (o_M, o_N)* will have the same velocity

* I have adopted the following notation: o_M denotes a point in M, and moving with it, momentarily coincident with the point in space o.

($\dot{\mu} \cdot om$ or $\dot{\nu} \cdot on$) and in the same direction. If, as the wheels revolve, the relative velocity $\dot{\mu}:\dot{\nu}$ varies, the point *o* will move, otherwise it will be stationary. Its locus in space is in the indefinite line of centers *mn*, it has also loci in *M* and *N*. If $\dot{\mu}:\dot{\nu}$ is constant these are circles called *pitch circles*, in general they may be called *pitch curves*. As *M* and *N* have the same velocity at the point *o*, and in the same direction the point *o* has the same velocity relative to *M* and *N*, and in the same direction.

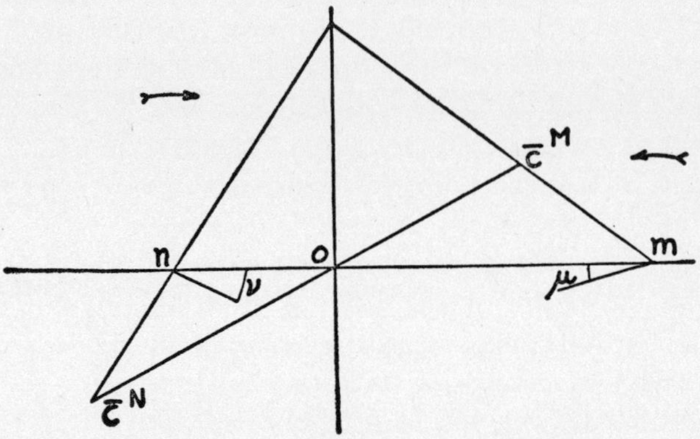

Fig. 1.

Hence, 1st the pitch curves are tangent at *o*, and 2d (the point of contact having the same velocity in both curves), they may be considered as rolling on each other without slipping. In fact *o* is simply the momentary center of the relative motion of *M* and *N*. It is evident, that if $d\nu/d\mu$ be given in terms of μ or ν, or in general if we have any equation, differential or not, between μ and ν, the pitch curves are determined, and again that any curve in *M* or *N* may be assumed as a pitch curve, and that it will determine the pitch curve in the other plane, and the relative velocity.

§3. The equations between *mo* and μ, *no* and ν, are evidently the equations of the two pitch curves referred to polar coordinates. Among the most simple and important analytical relations of these curves are the following, in which *r* is the radius vector (*om* or *on*) ψ = the angle included by the arc and rad. vect., S = the arc (measured from a corresponding point in the two curves), ρ = radius of curvature.

$$r^M + r^N = \text{const.} = mn^*$$

$$S^M = S^N$$

$$\psi^M + \psi^N = \pi$$

$$\frac{1}{\sin\psi}\left(\frac{1}{\rho^M}+\frac{1}{\rho^N}\right) = \frac{1}{r^M}+\frac{1}{r^N}\cdot$$

Also if $\bar{c}^M$ and $\bar{c}^N$ be the centers of curvature of the pitch curves, and $m\bar{c}^M$, $n\bar{c}^N$ be joined these lines will intersect in a line drawn through o perpendicular to mn. This is a simple case of the general proposition demonstrated in §8. The preceding equation containing ρ^M and ρ^N may easily be derived from the following note, if it is observed that m is the center of curvature of the path of n in M.

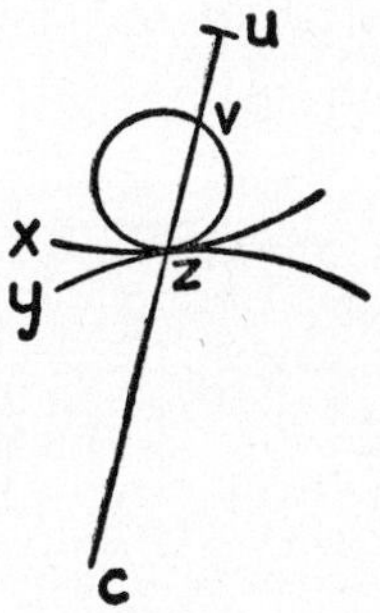

Fig. 2.

Note. The following propositions may often be of service in connection with this subject, yet seem hardly entitled to a demonstration here.

Let the curve x roll upon y at z carrying the point u, the curvatures being ρ^x and ρ^y† (positive when they are convex toward each other) let c be the center of curvature of the path of u, and γ the angle contained by uz and the normal.

$$\left(\frac{1}{uz}+\frac{1}{cz}\right)\cos\gamma = 1/\rho^x + 1/\rho^y$$

where uz is positive, when u is on the same side of z as x is of y and cz positive when c is on the same side of z as y is of x.

Or, draw a circle of which the

$$\text{diameter} = \frac{1}{\dfrac{1}{\rho^x}+\dfrac{1}{\rho^y}},$$

* The notation r^M denotes that the line etc. r is in M, but does not imply any idea of momentary coincidence as o_M does.

† [The author evidently intended $1/\rho^x$ and $1/\rho^y$. E.O.W.]

tangent to the curves at *z*, and on the same side of *z* as *x* is of *y* which cuts *uz* at *v*, the point *c* may be found by the proportion $uv:uz::uz:uc$ attention being paid to the signs of the segments. This is immediately deducible from the preceding proposition.

RELATIONS BETWEEN THE PITCH CURVES AND THE CURVES OF ACTION

§4. Let *M* drive *N*, the point of contact of the wheels being at the point *a*, which I will call the point of action. The veloci-

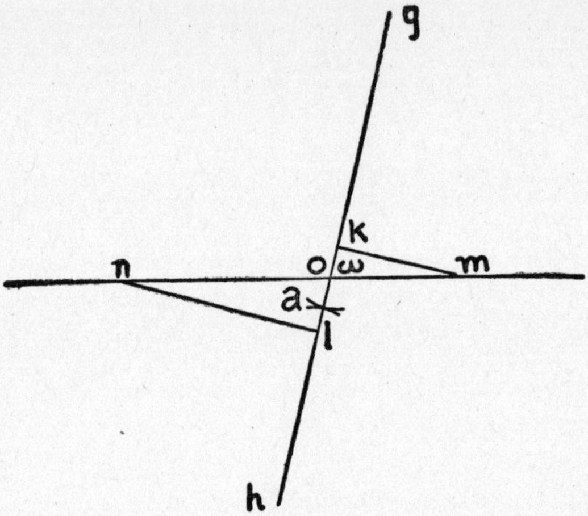

Fig. 3.

ties of the parts in contact, resolved along the common normal (to the surfaces in contact) must be equal. Draw the common normal *gh*, and perpendicular to it from the centers, *mk*, *nl*. a_M and k_M have the same velocity along *gh*, on account of their invariable distance, as also a_N, l_N; hence the above relation gives us $\dot{\mu}\cdot km = \dot{\nu} ln$, or $\dot{\mu}:\dot{\nu}::ln:km$, or $\dot{\mu}:\dot{\nu}::on:om$. By comparing this with §2 we see that *o* is the point of contact of the pitch curves, and we have the following general proposition.

The common normal to the acting surfaces passes through the point of contact of the pitch curves. This normal I will call the *line of action*. The locus of the point of action in space I will call the *locus of action*. The acting curves are evidently the loci of the same point in M and N. I will call these S^M and S^N, and their centers of curvature c^M, c^N.

§5. It is evident from the doctrine of evolutes that the normal *ag* may be considered as rolling upon the evolute of S^M, and *ah*

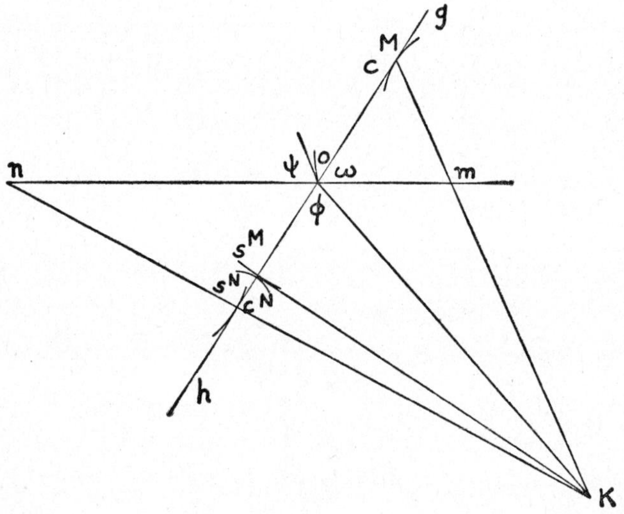

Fig. 4.

[The author's drawing of Fig. 4 carries this comment: "The perpendicularity of *ka* to *ao* in this fig. is accidental. *ka* is perpendicular at *a* to the locus of action (not represented)." E.O.W.]

upon that of S^N. Hence *gh* may be considered as rolling upon the two evolutes of the acting curves: and

The velocity transmitted by any two curves (as acting curves) is the same as if their evolutes acted by means of a line rolling on them.

§6. If we take any point a' in the line *hg*, its loci as the wheels move, S'^M, S'^N, are capable of transmitting the same velocity, for *gh* will roll upon the evolutes as before. This is simply the principle well known in practice, and sufficiently evident of itself, that if for one of two acting curves we substitute another at a constant normal distance from it, and for the other, another having the same constant normal distance from it, but on the other side of it, the motion transmitted by the two curves thus formed will be the same as that transmitted by the original curves. There is however this limitation,—that the distance aa' must not include a center of curvature of the acting surfaces, as will be apparent from the following principle.

§7. It is to be observed that if the point of action is between the centers of curvature, the curve having its center behind must be the driver, otherwise the curve having its center most in advance.

§8. Draw mc^M and nc^N, the intersection of these lines (k) is the center of motion of the normal gh. Draw ka; it must be at right angles to the locus of action. Draw ko; it must make the same angle with the normal, as the line of centers makes with the pitch curves. To prove this, call these angles ϕ and ψ,—the inclination of the normal to the line of centers, ω; and the velocity of either wheel at o, v. The velocity of o along mn is $v \cot \psi$. The velocity of o_{gh} along gh is $v \sin \omega$, as it is the same as that of a_{gh} or a_M in the same direction. See §4. The actual motion of o_{gh} is perpendicular to ok, hence its motion perpendicular to gh is $v \sin \omega \cot \phi$. Now the motion of o, the intersection of gh and mn, must be equal to the velocity of gh perpendicular to itself at o (i.e., of o_{gh} perpendicular to gh) $\times$ cosec ω. Equating these values we have $v \cot \psi = v \sin \omega \cot \phi$ cosec ω, $\therefore \psi = \phi$.

§9. By these principles, if we have chosen any locus in which we wish the point of action to move, the pitch curves being supposed known, we can find the centers of curvature for any point of the acting curves by a simple geometrical construction. From a draw ak perpendicular to the locus of action,—from o, ok making $\phi = \psi$. The intersection of these lines gives us k. Join km, kn,—the intersections of these lines with gh give us c^M, c^N.

§10. Again, the pitch curves being known, if either of the

points c^M or c^N be given, we may find the other (the locus of action and the point a being indetermined), by one of the propositions given in the note to §3. For one may easily convince himself that c^M is the center of curvature of the path of c^N in M, or vice versa.

But the most important theorem on this subject is one which I shall now proceed to give,—a theorem long honored as fundamental in this science.*

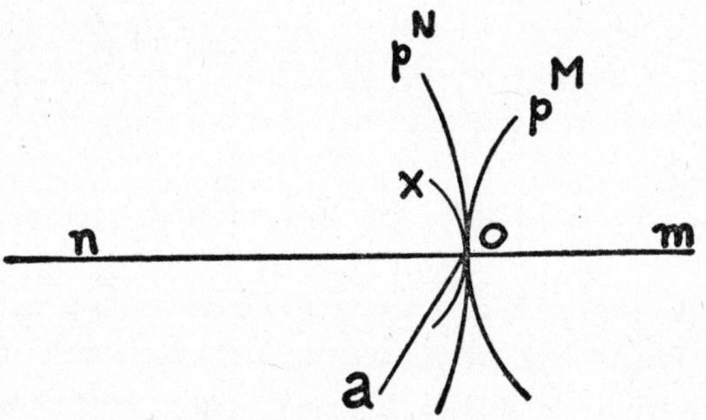

Fig. 5.

§11. If as the pitch curves roll together, any other curve rolls on them at their point of contact, carrying a point connected with it, this point will describe relatively to the two wheels two curves, which will be continually tangent to each other and whose normal from the point of contact will continually pass through the point of contact of the pitch curves.†

§12. If p^M and p^N are the pitch curves and x the other curve with which the point a is immovably connected, it is evident that as the wheels turn (the pitch curves rolling) the point a is always describing in M and N curves perpendicular to ao, and consequently tangent at a.

§13. It would seem however to require more demonstration

* See "Blake on the form of teeth of cog-wheels," Am. Jour. Sci. 1824, and Mr. Airy "on the forms of the teeth of wheels." Camb. Phil. Trans. 1827.

† It will be convenient to call such curves *conjugate curves*, the term conjugate curves will therefore imply that the pitch curves are supposed determined.

that the converse of this proposition is true,—that all conjugate curves can be thus described. To prove this, let *a* be the varying point of contact of the conjugate curves. Join *ao*. We may imagine a plane *A* attached to *a* and revolving about it with such velocity that o_A and o_M have the same velocity perpendicular to *oa*. But o_A and o_M have the same velocity along *oa*, viz. (from invariable distances) the velocities of *a* and a_M (resolved in the same direction), which are evidently equal. Hence o_A and o_M and (by a similar process) o_N have the same velocity and in the same direction, and *o* traces in *A*, *M* and *N* three curves which roll together at *o*, as may be proved by reasoning analogous to that of §2. Two of these curves are the pitch curves, the other is a curve connected with *a*, and may be considered to guide the motion of *a* by rolling upon the pitch curves while *a* describes the conjugate curves.*

§14. We will next inquire whether all conjugate curves are available for acting curves. Let S^M and S^N be such curves and *gh* the common normal, S^M being on the side of S^N which is toward *g*. If the motion is to be toward *h*, S^M can push S^N in that direction, or rather if S^M and S^N be made the bounding lines of teeth, S^M having its tooth towards *g*, and S^N towards *h*, the *M* tooth can drive the *N* tooth. If the motion is in the opposite direction, the *N* tooth can drive the *M* tooth. But in the first case *N* cannot drive *M*, nor in the last *M* drive *N*. In general then all conjugate curves can be made to transmit motion according to the pitch curves, but in only two of the

* The principle enunciated above is important, as it always affords a simple mechanical method for describing acting curves, provided that the pitch curves can be formed mechanically. In many cases, however, the most desirable forms of teeth require the acting curves to cross the pitch curves at an oblique angle, in which case the generating curve must make an infinite number of revolutions, its form approximating to that of an equiangular spiral near its pole, and this principle becomes inapplicable as a *mechanical* solution. Of course it is not necessary in practice that the generating curve be rolled on the pitch curves simultaneously, but care must be taken to give it initial positions upon the pitch curves, which it might have if it were rolled on the two simultaneously.

four supposable ways.* The direction of the motion being given the curve behind can drive the curve before, but not vice versa. (That direction along the normal is called forward, toward which the point of action moves.) This rule suggests its own exceptions, viz., the intermediate cases: *1st*, when the point of action does not move toward either end of the normal. In this case the point of action must be in the line of centers and the line of action coincides with the line of centers, making the intersection of these lines indefinite and the velocity indeter-

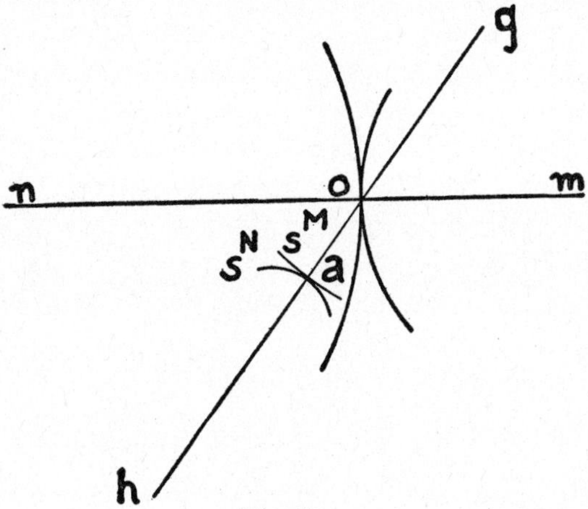

Fig. 6.

mined. *2d*, when the curves cross each other. If the curves have the same curvature in the same direction at the point of contact, they will in general cross each other there and of course be entirely unfit for use.† The same is true if one curve forms a

* It does not follow that teeth bounded by conjugate curves can transmit motion according to the pitch curves, for the additional condition is necessary that the teeth be on the right side of the curves.

† Hence the involutes of pitch circles are entirely inadmissible. There is a curious exception to this, when one circle becomes infinite (a rack). The involute then becomes a straight line, which is conjugate to (or rather of which one point is the conjugate of) the involutes of the other pitch circle. Yet from the general case we should expect that its curvature would be equal to that

cusp at the point of action. But if the point of action is not in the line of centers, and the conjugate curves do not have the same curvature at the point of action, in the same direction, nor either form a cusp there, the parts of the curves immediately adjacent to the point of action at least will always be capable of transmitting motion according to the pitch curves in two* of the four supposable ways. They are not however, necessarily eligible for continuous motion, as one of them may come in collision with the other or some other part of the other wheel in another part of its revolution.

§15. It will be convenient to call the curve and point, which by rolling on the pitch curves trace the curves of action, the *graphic curve* and *point*.

It is evident that as the graphic curve rolls on a pitch curve it must not change from one side of the pitch curve to the other (in regard to the parts adjacent to the point of contact), except when the graphic point coincides with the point of contact, or a cusp would be formed. But when the graphic point comes up to the point of contact, *then* the graphic curve *must* change sides, otherwise a cusp will be formed. Hence, we may conclude that if, as the graphic curve rolls simultaneously upon the two pitch curves, it is at any time on the same side of both, it must continue so, until the graphic point (the point of action) comes to the point of contact of the pitch curves, and then it must change to the other side of the two.

But if the graphic curve is intermediate in curvature between the pitch curves so as to fall between them, it must continue so until the graphic point comes to the point of contact, when the description of the curves must cease, for the graphic curve

---

of the other involute at the point of action. The explanation seems to be that although the involute of an infinite circle may be considered to have a curvature initially, indeed an infinite curvature at the very origin, yet this curvature is only for a portion of the line infinitely short (the infinite curvature is confined to a distance which may be regarded as an infinitesimal of the third degree), hence it does not produce any (finite) change in the direction of the line, which therefore we may again consider as straight.

* The influence of friction is here disregarded.

cannot change sides of both curves at once. That the curves just generated (by a graphic curve between the pitch curves) are unfit for use when the pitch curves are convex toward each other I will now proceed to show. Let p^M and p^N be two pitch curves in position, convex toward each other, and let the graphic curve Γ which falls *between* them have generated by the graphic point a the acting curves S^M, S^N. These are evidently perpendicular at a to the pitch curves, and convex as in the figure. Now if the motion be continued in the direction of the arrows, S^M and S^N must evidently come in collision, and are therefore inadmissible.

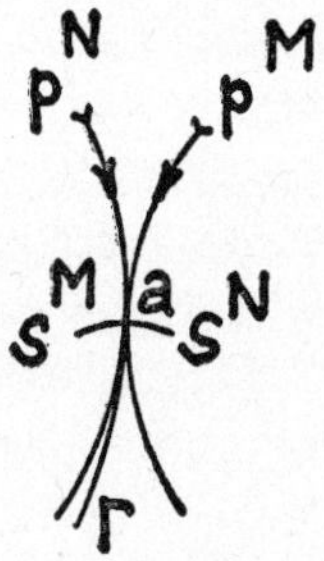

Fig. 7.*

§16. The question may arise: has every curve a conjugate? This reduces to another: can every curve be produced by the revolution of some curve upon the pitch curve? Let S^M be the curve and p^M the pitch curve,—draw a normal to S^M at any point a cutting p^M at o. Now suppose a plane A attached to a, and a to move along S^M, o will therefore move. o_A can have no velocity toward a, and as the revolution of A about a is yet indetermined, we may make the condition that o_A shall have no velocity perpendicular to oa. Therefore o_A has no velocity at all and its locus in A is a curve rolling upon p^M. It is evident from the above that if the perpendiculars to a curve do not cut the pitch curve, it has no conjugate. If they do, it will have a number of conjugates corresponding to the number of intersections of the normals with the pitch curve. I will remark, that it

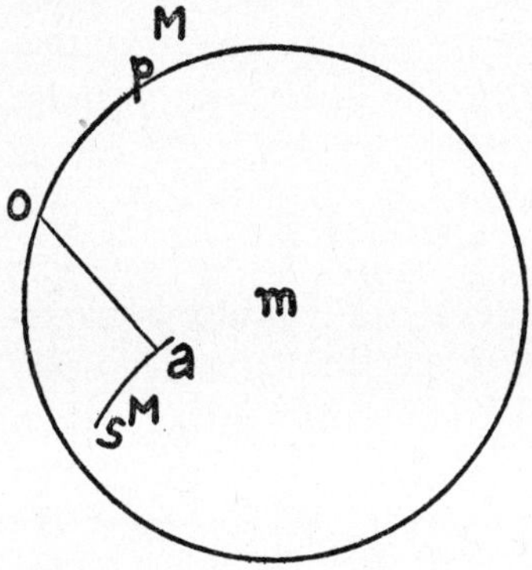

Fig. 8.

[* Designated Fig. 8 in the MS. Figs. 7, 10, 11, and 12 have also been renumbered in editing to preserve a consecutive order. E.O.W.]

may be the case that the curve is suitable for driving one of its conjugates, and being driven by the other, as a radial line, or that it is, for the moment suitable for driving both its conjugates (as in the case of the part of an involute of a circle less than a pitch circle, which is within the pitch circle), the direction of the motion being the same.

§17. As the pitch curves are always conjugates, and therefore in general capable of transmitting motion in two of the four supposable ways, it will be well to determine the cases in which they can be used. As o_M and o_N leave the line of centers, they move in a line perpendicular to *om* and with the same velocity.

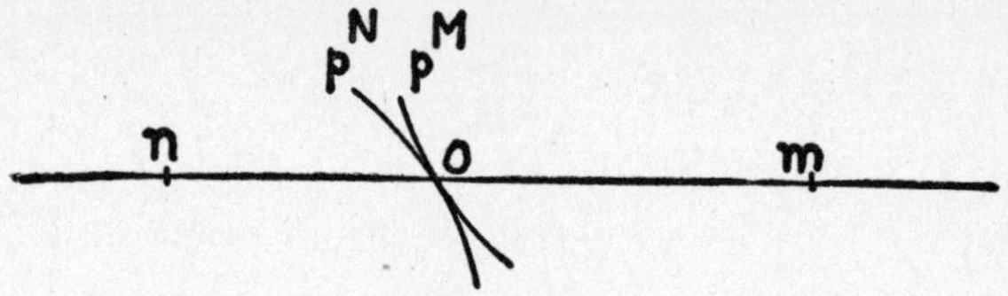

Fig. 9.

But o_M curves toward *m*, and o_N toward *n*, i.e., points in the pitch curves which were in contact in the line of centers, separate toward their respective centers of motion, and the relative position of the pitch curves must be as in the fig.—viz. p^M toward *m*, and p^N toward n. Whence one may easily conclude, that a wheel can drive another by the pitch curves, only when the relative velocity of the second is increasing.

§18. I will remark finally upon this part of the subject, that the relation between the conjugate curves is, that each is the envelop of all the *roulettes* described by the points of the other, or that each is the envelop of all the different positions of the other, in the plane revolving with the wheel of the first. This latter expression of the relation gives us a simple mechanical approximation method for finding a conjugate to any curve. See Willis, Princ. Mech. p. 71.

MODIFICATIONS OF FORCE INTRODUCED BY THE FRICTION OF THE ACTING SURFACES

§18½. It is evident that the actual pressure between M and N is in a line passing through a, making an angle with the normal equal to the angle of friction. Let it cut mn in l. If we call the moments of the pressure about m and n, P and $\mathfrak{Q}$, we have

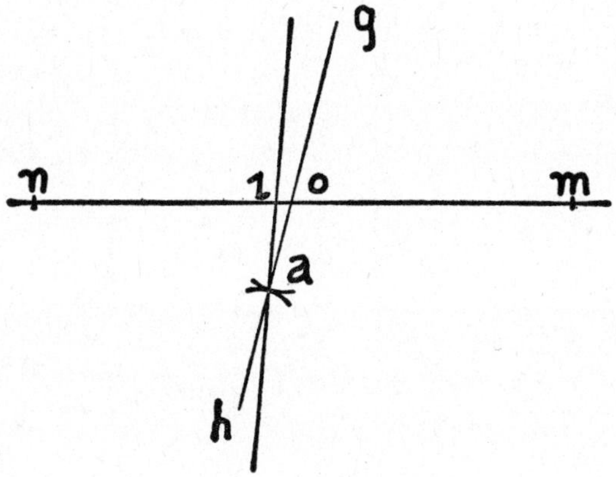

Fig. 10.

$p:q::ml:nl$. If there was no friction we would have $p':q'::mo:no$. ∴ making p constant we have for the modulus* of the system

$$\gamma = \frac{q}{q'} = \frac{nl \cdot mo}{ml \cdot no}$$

where no allowance is made for friction of the axes.

CASE OF CONSTANT RATIO OF VELOCITY

§19. Let us now limit ourselves to the consideration of the case most important in practice, the case when the velocity ratio is constant, and the revolution of the wheels in opposite directions. The pitch curves are then circles. As it is desirable

* The modulus of a machine is here taken to be the ratio of the work done to that which would be done if there was no friction, the power being constant.

that the friction should be as small as possible, we will assume that the point of contact of the pitch curves is one point of the locus of action. Hence (§15) the graphic curve must fall on the same side of both pitch circles, and the normal to the locus of action must cut the line of centers between the centers of the pitch circles, viz. at the center of curvature of the graphic curve. Hence the point of action recedes from one center and

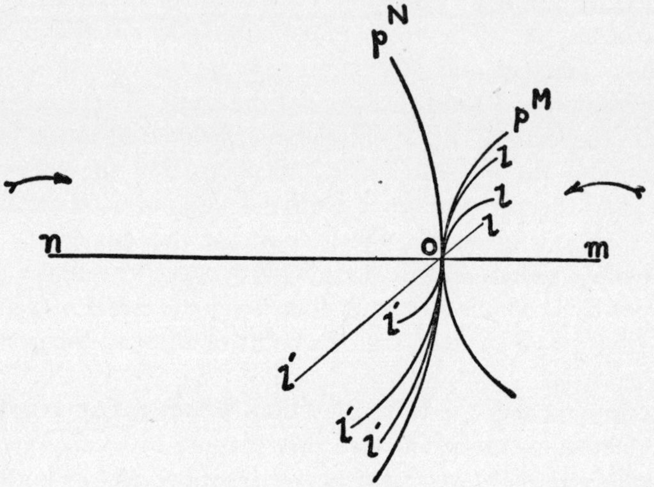

Fig. 11.

approaches the other, and if it passes through the point of contact of the pitch circles as there the graphic curve changes sides in respect to the pitch circles, the point of action must recede from the same center and approach the same center. One may easily convince himself that it recedes from the driving center and approaches the driven center. Thus (Fig. 11) *l l′* is a possible locus of action, for *M* to drive *N* in the direction indicated in the fig. (or *N*, *M* in the reverse) because *l l′* continually recedes from *m* and approaches *n* (regarding *l* as the first point and *l′* as the last point of *l l′*, because if the motion is as indicated in the figure, the point of action moves from *l* to *l′*).

§20. As the dynamical relations of the wheels are immediately derived from the position of the point of action, in choosing from the infinite number of conjugate curves, those most suitable in any case, it is natural to choose first the locus of action, and from it derive the curves of action. I have already shown how to find points in the evolutes of the acting surfaces, the locus of action being given, §9. I will add another relation between the locus of action and the graphic curve. As this curve Γ always cuts *mn* at right angles, its center of curvature is always in *mn*. As o^{Γ} moves perpendicularly to *mn* the evolute of Γ rolls upon *mn*, and this rolling may be considered as guiding the motion of Γ, and therefore of the point of action, which moves in the locus of action. Therefore, the locus of action being given to find the graphic curve and point, and the curves of action; find the curve and point which rolling upon *mn* will describe the locus of action, the involute of this curve (namely *that* involute which will pass through *o*, when the curve is in some position, which it has in describing the locus of action) will be the graphic curve, and the point mentioned the graphic point.

§21. In choosing the locus of action, we must pay attention to the following condition, viz., that it must be of such length that while the point of action moves from one end of it to the other, the wheels may revolve a certain distance, which, measured on the circumference of either pitch circle, I will call the *stroke*.

Let S = the stroke regarded as variable
Let $r = ao$ (position when a is toward h)
Let $\omega = gom$
then $dr = dS \sin \omega$ (Compare §)*
or

$$\frac{dr}{\sin \omega} = dS$$

$$\int \frac{dr}{\sin \omega} = S.$$

[* Possibly the author intended to refer to §8, where the following sentence occurs: "The velocity of o_{gh} along gh is $v \sin \omega \cdots$." E.O.W.]

Hence the locus of action must be such that

$$\int \frac{dr}{\sin \omega}$$

shall have a certain value. (The integral is to be taken for the whole extent of the locus.)

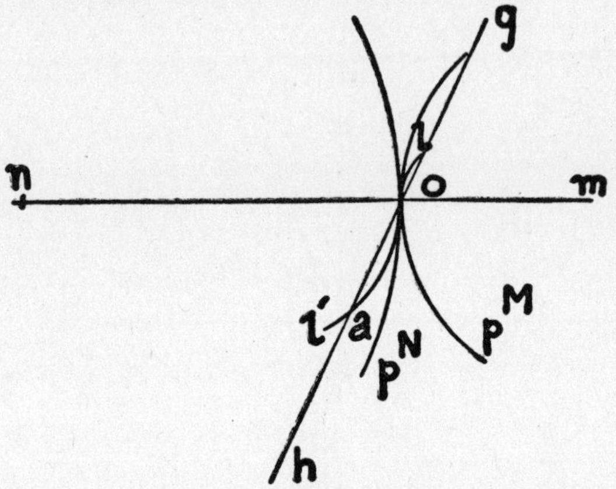

Fig. 12.

INVOLUTE TEETH

§22. The principles developed in §5, naturally suggest the involutes of circles as the acting curves. Thus the line of action *gh* rolling upon the circles *d*, *e*, the point *a* will describe involutes of the same circles in *M* and *N*; the locus of action is a straight line, coinciding with the line of action. The radii of the circles *d*, *e*, have evidently the same ratio as the radii of the pitch circles.

§23. As their ratio to the pitch circles is immaterial, involutes of the same circles will answer for different distances of centers, i.e., if we have two wheels with involute teeth which will work together, the distance of the centers is immaterial (within certain limits, which depend on the length of the teeth). This property of the involute is a very important one in practice, owing to the difficulty of perfect adjustment. It will be well to examine whether any other form of teeth possesses it. If the driver move with constant velocity, and the distance between the centers be increased by a distance δx, the driven wheel will fall back a distance δv. If the velocity of the driven wheel is to be constant after the change, δv must be the same for all positions of the wheels. Let $gom = \omega$. If the distance of the centers

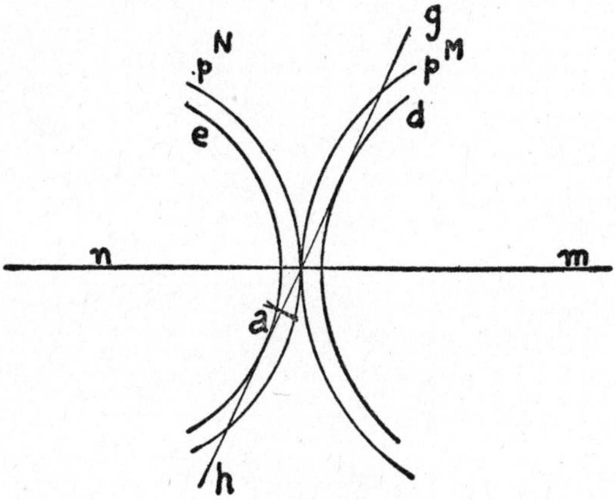

Fig. 13.

be increased by an infinitesimal δx the acting curves will be separated to the amount of $\delta x \cos \omega$ and the driven wheel will fall back in angular distance $\delta x \cos \omega \cdot no \sin \omega$.* If this is to be independent of the primitive positions of the wheels, ω must be constant and the acting surfaces are involutes of circles.

§24. Draw *oy* perpendicular to *mn* at *o*, making it equal to the desired stroke which is to commence when the point of action is in the line of centers. Describe a semicircle upon *oy*, it is the limit of all straight loci of action, of which the stroke is equal to *oy* (§21). Again describe on *no* a semicircle, *nzo*, it forms the extreme possible limit of all straight loci of action. See §19. In fact beyond this limit the involute in *N* will form a

[* The author must have intended a sign of division, rather than multiplication, between $\cos \omega$ and *no*. E.O.W.]

cusp. See Fig. 13. Lay off an arc *ou*, equal to double the angle of friction, ϵ. The line of pressure for the extreme point of any straight locus of action of stroke *oy* will pass through this point *u*. It is noticeable that the friction may be decreased without limit by making *l* approach *o*, along the arc. The obliquity of the line of action would cause the pressure upon the axes to increase,—*finally* by the multiplier $1/\sin \epsilon$ nearly. For this and other reasons this form of teeth is of no practical value, I

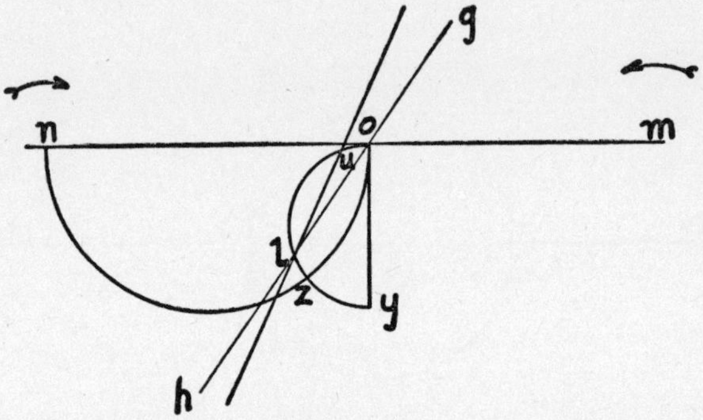

Fig. 14.

merely give it to show that there is no proper minimum to the friction at the point of contact. From the same construction we see that there is no limit to the stroke. We may make *oy* as large as we choose and taking the point *l* between *z* and *o*, and joining *lo*, we have a possible locus of action of the proper stroke. Hence there is no limit in regard to the number of the teeth,—a wheel of one tooth may drive a wheel of one tooth. All this however *only* when the motion is in only one direction, and the same wheel always drives.

EPICYCLOIDAL AND HYPOCYCLOIDAL TEETH

§25. The form of teeth most naturally suggested by the principles of §11, is that of epicycloids and hypocycloids. If a circle $\Gamma'\Gamma''$ rolls simultaneously upon p^M and p^N, any point *a* in it will describe in *M* and *N* conjugate epi- and hypo-cycloids.

The locus of action is evidently a circle coinciding with $\Gamma'\Gamma''$. The part of this locus suitable for *M* to drive *N* is $o\Gamma''x$. By using another graphic circle on the other side of the pitch curves, we may add to this locus a part of $o\Gamma'''y$. These circles must not fall between the pitch circles. The limiting case when the graphic circle coincides with a pitch circle is not ex-

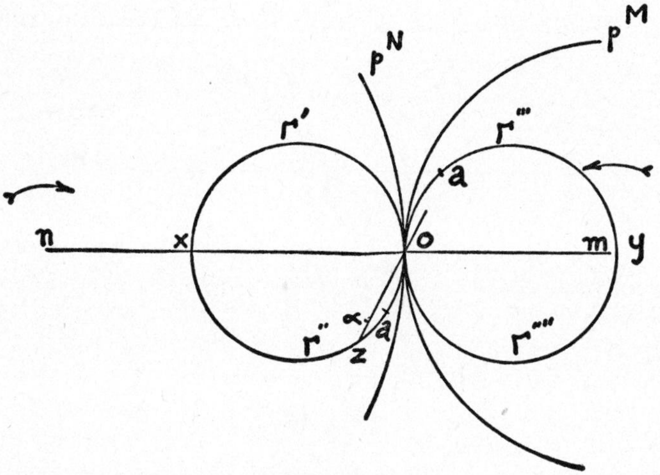

Fig. 15.

cluded. We then have a point and epicycloid for acting curves. Of course an angular point is undesirable in practice. We may approximate to the same action, however, by taking a graphic circle a little less than the pitch circle. Or, we may strike a small circle about the point, and in the other wheel lay off a curve having a constant normal distance to the epicycloid (§6). In this and all cases where curves of constant normal distance from epi- and hypo-cycloids are taken, we may easily find the locus of action by supposing one end of a line (*z*) to move along the graphic circle while the line slides through the point *o*. Lay off $z\alpha$ equal to the constant normal distance, and α will describe the locus of action.

§26. A common method is to take graphic circles of radius equal to one-half of the radius of the nearer pitch circle. By this means the curves within the pitch circles in both wheels become radial lines.

§27. The stroke of any locus of action for epi- and hypocycloidal teeth may be derived from §21 but it is more easily found from the manner in which it is described. It is equal to the length of the locus (measured in the arc).

§28. Theoretically, it is more simple to consider the graphic curve as a single curve made up of two circular arcs, of reversed curvature, with the graphic point at the point of junction.

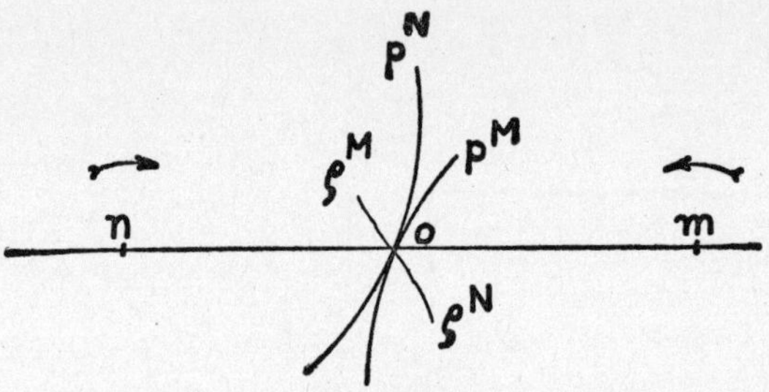

Fig. 16.

VELOCITY RATIO VARIABLE

§29. Let M drive N in the direction indicated in the figure, o (the point of contact of the pitch curves) moving toward m (the center of the driving wheel). Assuming that the acting curves meet the pitch curves, let the wheels at this moment be so placed that they meet them at o. As the wheels turn the point o_N describes in M a roulette ς^M, beyond which the tooth in M must not extend (to avoid interference with o_N). This (ς^M) is a sort of limit to all the admissible curves of action in M, and will evidently give a less oblique line of action than any other. Its conjugate is of course the point o_N. In like manner in the revolution of the wheel o_M describes in N a roulette ς^N which of all admissible curves of action gives the least oblique line of action. Therefore the point o_M and ς^M in M, and ς^N and o_N in N, as acting curves, give the least oblique line of action, and will generally be *theoretically* the best.

§30. Again let o move toward n. If we take for a graphic curve a spiral which makes an angle with its radius vector equal to the complement of that which either pitch curve makes with its radius vector, and for the graphic point the pole of the said spiral, we may make the line of action always perpendicular to the line of centers. Or, if the motion is always in the same direction, we may by similar means make the line of action inclined at a constant angle ϵ, so as to make the line of pressure always perpendicular to the line of centers.

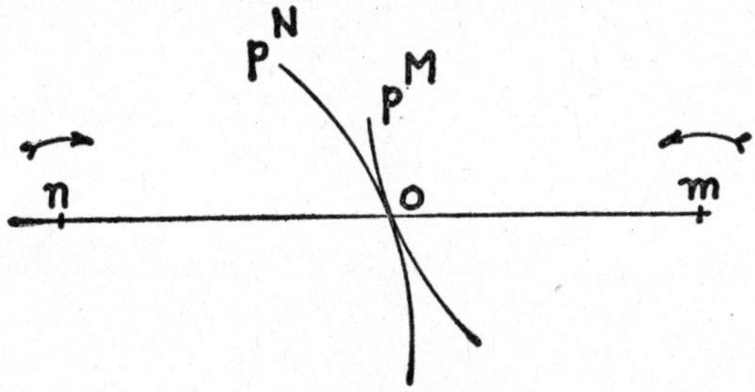

Fig. 17.

In this case we may make teeth of very satisfactory action, but the graphic curve does not afford a means of mechanical description. It will be well, therefore, to develop the subject a little further, independently of the graphic curve.

§31. To the necessary condition that the line of action shall pass through the point of contact of the pitch curves, let us add the condition that it shall be perpendicular to the line of centers. From o draw ox making $gox = p^{N \text{ or } M}om$, the center of motion of gh is in ox at infinite distance. Draw parallel to ox, mc^M, nc^N,—c^M and c^N are the centers of curvature of the acting curves, or the points where gh may be considered as rolling upon the evolutes of the acting curves e^M, e^N. It will be convenient to have a name for these evolutes (of acting curves of which the line of action is perpendicular to the line of centers). For want of a better, I will call them the principal evolutes. Let us examine their nature. Let us refer them as we did the pitch curves (§3) to the centers of motion (m, n,) as poles.

§32. It is evident that the radius vector (o, m) of the pitch curve is the perpendicular upon the tangent from the pole, for the corresponding point of the prin. ev.

Again the angles mop^M, mce^M, are equal, i.e., the angle made by the pitch curve with its rad. vect. is the same as that made by the prin. ev. with its rad. vect.

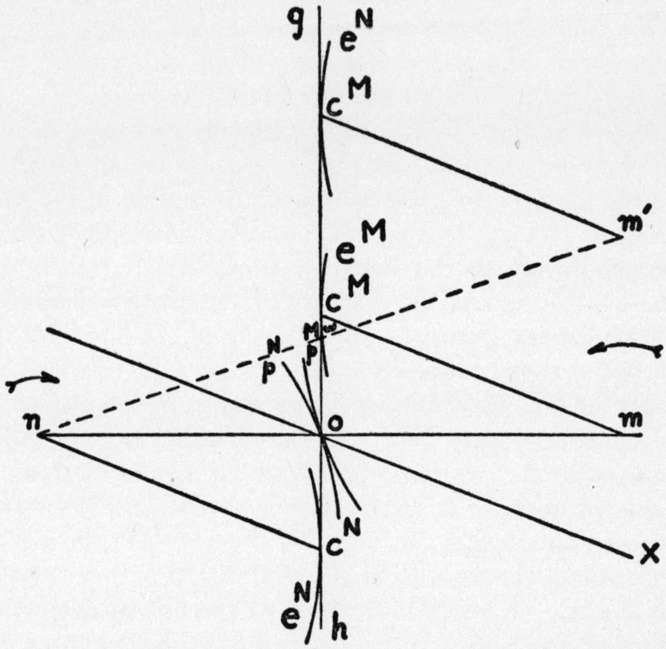

Fig. 18.

Again the tangent of the evolute is always at right angles with the rad. vect. of the pitch curve, so that the angle representing the direction of tangent of the prin. ev. is the same as the angle representing the direction of the rad. vect. in the pitch curve (these angles being measured from lines of reference fixed in M perpendicular to each other).

If therefore, we have found any relations between any of the six quantities, the rad. vect. the angle at the pole, the angle made by the curve with the rad. vect.; of the two pitch curves;

the same relations will hold between the corresponding quantities of the following six, viz., the perpendicular from the pole upon the tangent, the angle made by the tangent with a fixed line, the angle made by the curve with the rad. vect.; of the two prin. ev.

§33. In a similar manner we might investigate the nature of the evolutes of acting curves of which the line of action makes with the line of centers any other constant angle. We may, however, save ourselves the trouble of an independent investigation by the following considerations.

It is evident that if M were placed with its center at m' instead of at m, the line mm' being perpendicular to nm, e^M would roll upon gh precisely as before, the motion of the wheels suffering no change. In this case however gh would no longer be perpendicular to the line of centers nm' but inclined at an angle ω. Moreover, the distance of the centers is increased by the multiplier $1/\sin\omega$, and relatively to the line of centers, which has changed its position, the evolutes have been turned an angle $90° - \omega$. Hence from the general principles of proportional figures, we may conclude, that if in M and N (in their original positions) we describe about m, n, curves similar to the prin. ev. but on a scale reduced by the multiplier $\sin\omega$, and turned about m, n, an angle $90° - \omega$ (both in the same direction), these curves will be the evolutes of acting curves, of which the line of action is inclined to the line of centers at a constant angle ω (being turned from the perpendicular an angle $90° - \omega$, in the *same direction* as the reduced evolutes were turned).

§34. The principal object of our present investigations is to enable us to give to the line of action, and thus to the line of pressure, that direction, which the dynamical relations of the wheels render most desirable, during that part of the revolution in which this direction is not limited by the condition of the teeth clearing one another. We may, however, from the principles just given form teeth for the whole perimeters of the wheels.

It is evident that in Fig. 18 the relative positions of c^M and

c^N in gh are such, that M can drive N in the direction indicated, but N cannot drive M in that direction, i.e., the wheel must drive of which the relative velocity is decreasing. If, however, we turn gh about o, and therefore ox, mc^M and nc^N moving so as to be parallel to ox—as gh passes the tangent to the pitch curves c^M and c^N will pass one another in gh, and a little beyond this position N can drive M, i.e., the wheel can drive of which the relative velocity is increasing. By taking a line of action of proper obliquity therefore, we can form teeth for any part of the perimeter of the wheels. When the relative velocity of the driver increasing, the action will not be so satisfactory as in the reverse case, yet it will be but little inferior to that of the teeth mentioned in §29. If we should adopt for the whole revolution a line of action a little more oblique than the greatest obliquity of the pitch curves to their radii vectores, we would have a single evolute for all the faces of the teeth which act in the same direction, another evolute would be necessary for the reverse faces. Such a construction, however, would have little but simplicity to recommend it.

§35. In cases in which the precise relation between μ and ν (the angles determining the positions of M and N) is unimportant, it will often be convenient to use the following construction. Take two convex figures in M and N, such that by means of a common tangent rolling upon them they will transmit a desired motion. These will be the evolutes of acting curves. These figures may easily be so taken that the common tangent (line of action) shall have a direction nearly as desirable as the case admits of. It is important to notice, that with teeth thus formed the smoothness of action is not affected by a limited change in the distance of the centers. A combination of a few arcs of circles may be made to answer the purpose very well. Or we may find by calculation or mechanical processes the general form of the evolute which will give a satisfactory action, and cut it out without particular regard to accuracy, for the smoothness of action is not at all dependent upon the particular form of the evolute.

If simplicity is the chief requisite, two eccentric circles may be taken, in fact, if the eccentricity be small this is a very good form. So also if one wheel is to revolve more than once during the revolution of the other, we may take one wheel eccentric and the other not so. Rectilinear polygons would give teeth bounded by circular arcs or combinations of the same.

The best form for these figures, is generally the principal evolutes for a part of the wheel and curves drawn a little within the pitch curves for the rest.

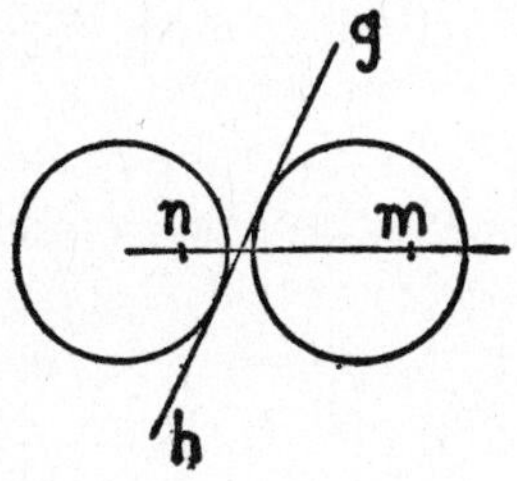

Fig. 19.

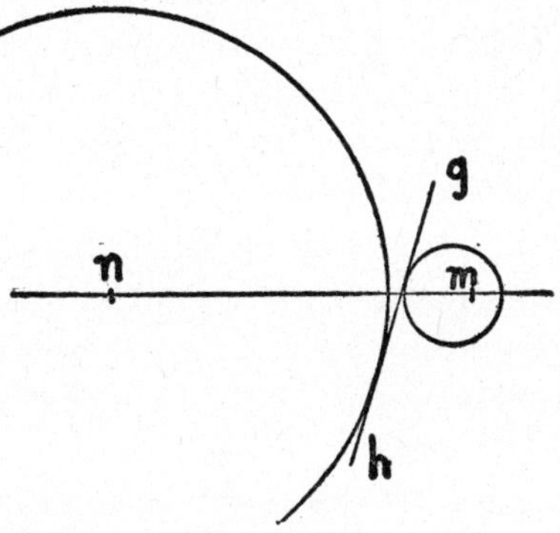

Fig. 20.

TEETH OF CONSTANT PRESSURE

§36. It is often desirable to give such a form to the teeth of wheels, that the driven wheel shall be impelled by a constant force, which shall not be affected either by the friction of the acting surfaces or by that of the axes, the force applied to the driver being supposed constant. It is evident that this condition is entirely satisfied if the line of pressure does not change its position. Let us examine the results of this supposition.

§37. Let *ij* be the invariable line of pressure, M the driving wheel, the motion being in the direction indicated in the figure. The locus of action is of course in *ij*. From *a* the point of action draw *ux* perpendicular to *ij*; the center of motion of the line of action is in this line (§8) and, as the line of action has an invariable direction, at infinite distance. Through a draw the line of action gh making $iag = \epsilon$.

Draw lines through m and n parallel to ux cutting ij in l and l', gh in c^M and c^N. These latter points (§8) are the points where the evolutes of the acting curves may be regarded as rolling on gh. The angles gc^Mm, nc^Nh are evidently equal to the complement of ϵ. Hence the evolutes of the acting curves are equiangular spirals, the constant angle being the complement

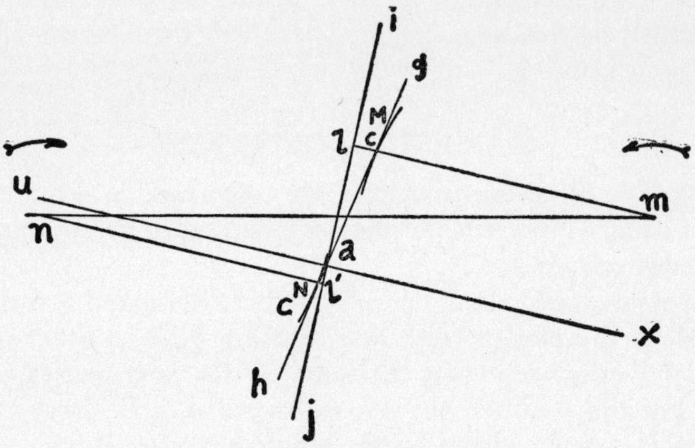

Fig. 21.

of the angle of friction. Again as xag = the angles made by the pitch curves with the line of centers (§8), the pitch curves are also equiangular spirals of the same angle.

§38. When the point of action is on the other side of the line of centers, the line of action takes a different direction, and the evolutes of the acting curves and the pitch curves are equiangular spirals of the same angle in a reversed position. It is evident that the locus of action is limited by the points l, l', as beyond these we have another branch of the involute for the acting curve, making a cusp with the first. This distance, ll', may be increased by increasing the obliquity of the line of pressure.

§39. If the action is only after the line of centers we may make the line of pressure as nearly perpendicular to the line of centers as will give us sufficient *stroke.* If the action is on both sides of the line of centers, there will be an angle $(180° - 2\epsilon)$ in the acting curves, and the line of pressure must be inclined to the line of centers at an angle not exceeding $90° - 2\epsilon$, in order that the acting curves may clear each other at the angular points. It may be desirable to take the locus of action so that the lines of action for its extreme points will cut the line of centers in the same point. There will then be no sudden change in velocity.

SUCCESSION OF TEETH

§40. The following principles are important in some cases.

Let μ and ν represent the angular positions of the driving and driven wheels, M and N.

Let $\nu = \phi(\mu)$ represent the relation between μ and ν as determined by the pair of teeth a, α, if there were no other teeth, and let $\bar{\nu} = \bar{\phi}(\mu)$ represent the same for the next pair of teeth, b, β. For the position where b, β succeed a, α we have $\nu' = \bar{\nu}'$. Before this point ν must be greater than $\bar{\nu}$, after it $\bar{\nu} > \nu$

$$\therefore \quad \frac{d\bar{\nu}'}{d\mu'} > \frac{d\nu'}{d\mu'}$$

or if

$$\frac{d\bar{\nu}'}{d\mu'} = \frac{d\nu'}{d\mu'}, \qquad \frac{d^2\bar{\nu}'}{d\mu^{2\prime}} = \frac{d^2\nu'}{d\mu^{2\prime}} \quad \text{and} \quad \frac{d^3\bar{\nu}'}{d\mu'^3} > \frac{d^3\nu'}{d\mu'^3}$$

etc., etc.

Provided, that the values of the differential coefficients are *continuous.*

FRICTIONLESS TEETH

§41. It seems to have been supposed by some writers that toothed wheels can be formed to work without friction at the acting surfaces, the teeth having the form of equiangular spirals or some other form.

It may easily be inferred from §17 that one wheel can drive another without friction only when the relative velocity of the driven wheel is increasing, and we have just seen (§40) that the velocity of the driven wheel cannot be diminished when one pair of teeth succeed another unless the coefficients of relative velocity for each pair of teeth become discontinuous at this point. This cannot be, although we make the acting (pitch)

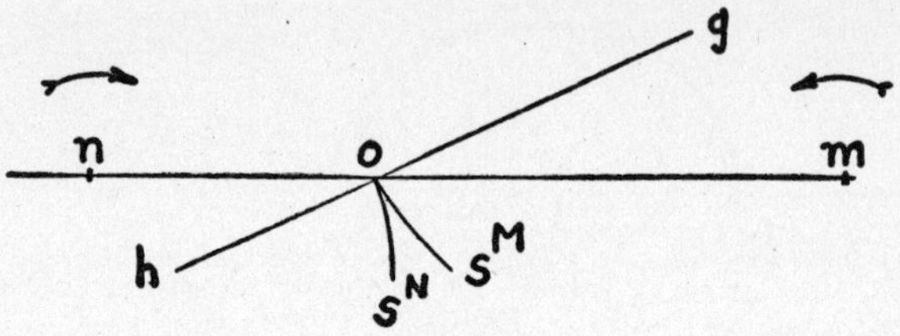

Fig. 22.

curves discontinuous, unless these curves cut the line of centers at right angles, at the moment of succession. Thus in Fig. 22 it is evident that if the acting curves be suddenly cut off, the point *o*, the intersection of the normal to the acting curves and the line of centers, which determines the relative velocity of the wheels, will not suddenly change its position, but move with finite velocity back toward *m*. So that these curves will continue to act and of course not without friction, even if some other curves are brought in which give *o* a position nearer *m*. If however, in the course of the motion, the acting curves s^M, s^N, of which the point of contact is in the line of centers, come to cut that line at right angles, *then* the continuity of $dv/d\mu$ is destroyed and the acting curves may be succeeded by another pair which also must cut the line of centers at right angles, in order that $d\bar{v}/d\mu$ may also be discontinuous. Unless both these differential coefficients are discontinuous, we will have, either just before or just after the time of succession,

$$\frac{d\bar{\nu}}{d\mu} < \frac{d\nu}{d\mu}$$

which makes the succession impossible.*

The only forms of acting curves which are admissible are evidently of a very objectionable character. I give an example in Fig. 23 of two two-toothed wheels when the acting curves are semi-ellipses. They are represented at the moment of succession. *M* is the driver for the direction represented, or *N* for the reverse direction.

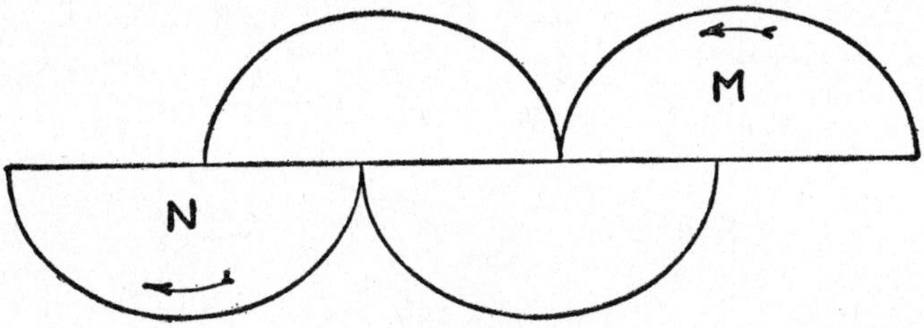

Fig. 23.

CIRCULAR ACTING CURVES

§42. If we take circular arcs for acting curves, the mechanical execution of the wheels will be rendered the most simple possible. In the choice of these arcs the greatest desideratum for general purposes is that there shall be no sudden change in velocity at the succession of the teeth, provided that the velocity ratio be moderately uniform, and the line of pressure does not deviate very far from the perpendicular. This requires that at the time of succession,

$$\frac{d\bar{\nu}}{d\mu} = \frac{d\nu}{d\mu}$$

and therefore that

$$\frac{d^2\bar{\nu}}{d\mu^2} = \frac{d^2\nu}{d\mu^2}$$

* In the first case we would have $\bar{\nu} > \nu$ before the succession, in the latter $\bar{\nu} < \nu$ after the succession.

and

$$\frac{d^3\bar{v}}{d\mu^3} > \frac{d^3v}{d\mu^3}.$$

If the two first conditions are satisfied, the latter will usually be found true, and we need only consider the two first. The first one requires that the line joining the centers of the first pair of arcs, and the line joining the centers of the second pair, shall cut the line of centers in the same point. The second may be satisfied (or the necessity of satisfying it evaded, according as we view the subject), by making a sudden change in the curva-

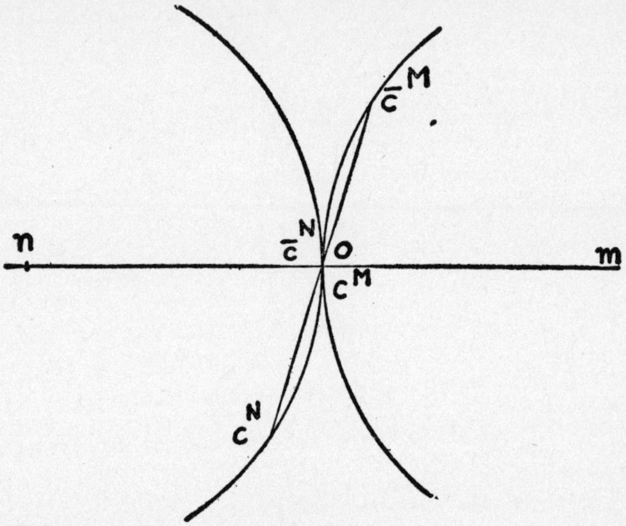

Fig. 24.

ture of the arcs at the point of contact for the time of succession, or by cutting them short at that point; otherwise the arcs must be so chosen that the velocity of the point *o* along the line of centers shall be the same for both pairs of arcs at the moment of succession, in this case of course

$$\frac{d^2\bar{v}}{d\mu^2} = \frac{d^2v}{d\mu^2}.$$

This, I believe, is always possible, although the practical calculation appears extremely complicated, except when the wheels are of equal size. Fig. 24 represents this case, where c^M and c^N are the centers of the first arcs, and $\bar{c}^M$ and $\bar{c}^N$ the centers of the second pair. That the velocity of *o* along *mn* is the same for the line joining c^M and c^N, and for that joining $\bar{c}^M$ and $\bar{c}^N$, is evident from sufficient reason, the figure being entirely symmetrical. If either wheel is to drive, it will be best to

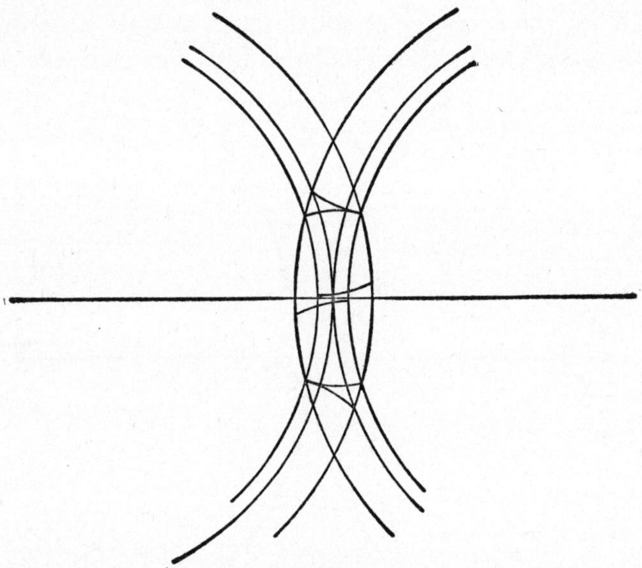

Fig. 25.

divide $c^M c^N$ equally for the radii of the acting arcs. It is to be observed that the variation in velocity is extremely small, and the friction and line of action about the same as in the best teeth of constant velocity-ratio. In Fig. 25, the acting arcs are drawn. I have cut them off in the figure where they are to cease to act, but this is not necessary; if they were continued further they would not interfere, which is the peculiarity of this construction.

COMMENTARY UPON THE GIBBS MONOGRAPH

"ON THE FORM OF THE TEETH OF WHEELS IN SPUR GEARING"

EVERETT OYLER WATERS

In his introduction, Dr. Wheeler has very aptly pointed out the contrast between Gibbs' early interests, as exemplified by the present volume, and his later devotion to the study of theoretical physics. While it is true that a degree of continuity may be claimed on the basis of his preference for the mechanical viewpoint throughout his career, it should also be remembered that, even in Europe, the line between pure and applied science had not been drawn until recently with anything like the sharpness of distinction that we recognize today. From Galileo to Coulomb and Poisson, the most eminent natural philosophers and mathematicians of the seventeenth and eighteenth centuries made frequent excursions into fields that are now of more interest to the civil and mechanical engineer than to the physicist.

As early as 1666, Hooke had observed that tooth contact in spur gearing, in order to be frictionless, must occur on the line of centers; Thomas Young, in his "Natural Philosophy," had demonstrated the properties of cycloidal curves, known at that time to be suitable for transmitting motion with uniform velocity ratio; Euler investigated the geometry of all suitable tooth profiles and laid down certain principles for approximating them in practice. According to Willis, he was the first to suggest the use of the involute profile, which today has practically supplanted all other curves that might appropriately be used. The British scientists, Brewster and Airy, who antedated Gibbs by more than a generation, wrote important monographs on the forms of gear teeth. In some cases, these studies had undoubtedly been prompted by a practical motive, such as the perfection of chronometer movements, or the construction of planetaria and the like; in others, it was nothing more than a mathematical recreation for the investigator. Euler's papers, for example, were considered so abstruse as to be worthless to the eighteenth or early nineteenth century engineer.

The present essay, although unsigned and undated, is unquestionably the work of Professor Gibbs. It is also, in all probability, his thesis for the doctorate in philosophy which was granted to him by Yale in 1863. This degree had been inaugurated only two years earlier, and its administration was still in the hands of the Yale College faculty (the Graduate School was not formally organized until 1892). The records of that time are confused and meager; there was no requirement that theses be published, and no complete file was kept in the Yale Library or elsewhere of theses that had been accepted. Shortly after Gibbs' death in 1903, his brother-in-law, Professor Addison Van Name, discovered the manuscript while making a search through some papers that Gibbs had kept in his study at his home. It had neither title page nor binding, but the style of handwriting indicated that it had been written at a much earlier date. Several years later, it was shown to Professors Beebe and Phillips, who, as Dr. Wheeler has stated, gave their formal opinion of its authenticity as a doctoral thesis. The writer, after reading the manuscript himself and noting certain informalities such as the crossing out of words, interlinear insertions, occasional omission of items in the figures which are referred to in the text, and a trifling number of lapses from technical accuracy, offers the suggestion that another, corrected, copy was submitted to the examiners for the degree, and that the present copy, in the nature of a final draft, was retained by the author for his own files.

In 1925, the manuscript was presented by his niece and nephews to the Yale University Library, where it was suitably bound and is now preserved in the Rare Book Room.

"On the Form of the Teeth of Wheels" is, in many respects,

quite unlike the typical modern dissertation in the field of science. To that extent, it reflects both the highly individual characteristics of the author, and the pioneer status of the doctorate in philosophy in this country in the mid-nineteenth century. There is no single outstanding objective to be proven by experiment or logical analysis of known data; no review of the existing literature is reported, although the author does make scant reference to Airy, Eli Whitney Blake, and Willis. It is a mere guess as to whether he was familiar with Camus' treatise, which was available in English and was reasonably well known at that time, or whether he had ever made a translation of Euler's papers in the *Acta Petropolitana.* However that may be, it is interesting to note that Gibbs' general method of approach, which is to define tooth profiles by their radii of curvature, is essentially that which was proposed by Euler.

It is also rather difficult to estimate the degree of scholarly background which Gibbs expected of his readers, as evidenced by his choice of propositions which did, or did not, merit a "demonstration" (i.e., proof). For example, a basic theorem on the evolute of a geometrical roulette is stated as a mere footnote to Section 3, whereas the theorem relating to the generation of tooth profiles by rolling a third curve on the given pitch curves is discussed most thoroughly in Section 11 and following; yet the latter had been demonstrated fairly recently by Airy, while the former is apparently supposed to be taken on faith, or accepted as a challenge of the reader's ingenuity in the geometry of infinitesimal displacements. In the main, however, the propositions follow each other in logical sequence, and, as one proceeds, he feels that he is gradually reaching the summit of an intellectual structure that is firmly founded and well joined. If he has a natural friendliness for the niceties of geometrical reasoning, he will be rewarded with a sense of satisfaction akin to that felt upon completing, say, a book of Euclid; if he is not so endowed, he had perhaps better not trouble himself with the austerities of style and extreme economy (one might almost say parsimony) in the use of words that characterize the entire work. Even the figures are devoid of many auxiliary lines that would help in a more rapid understanding of the demonstrations in the text.

While the subject matter is clearly in the field of plane geometry, there is ample scope for treatment by analytical methods. However, Gibbs very definitely eschewed this line of attack, and as a result the thesis contains a surprisingly small number of equations and formulas. The reader with a good mental grasp of spatial relations, who can picture exactly what goes on when three coincident planes are all moving at the same time, and a point in one is tracing curves in the other two, will have much easier going than the expert in algebraic manipulation. By the same token, whoever wishes to test for himself any of the author's propositions will find that the tools of classical geometry—a straight edge and compass—comprise all the equipment that he needs.

The thesis begins with an introductory paragraph, stating the practical utility of a study of gear tooth profiles and announcing the author's intention to confine himself to the particular case of spur gearing. From this point on, the subject is treated in eleven chapters or articles, a brief discussion of which may be of service to those who wish to gain some familiarity with the outlines of Gibbs' treatment before studying it in detail.

PITCH CURVES

In this article the author defines pitch curves (not necessarily circles) and establishes the well-known propositions regarding their tangency on the line of centers and the relative angular velocities of their respective planes. At this point, the reader would do well to fix in his own mind the geometrical background that Gibbs uses throughout, as a screen upon which to project his rolling curves, evolutes, and other constructions. It consists not of a single plane, but three, coincident and (except for certain points of mutual constraint) independent. The first is the plane of reference; the second, the plane of one of the pitch curves, denoted by *M*; the third, the plane of the other pitch curve, *N*. *M* is pinned to the reference plane at *m*, and *N* is likewise pinned at *n*, making *mn* the line of centers, which by the first equation of Section 3 has constant length. Occasionally an additional plane or two may be superposed, but the basic three are always assumed by the author and must be visualized by the reader, particularly in studying the illustrative figures.

The first division of the thesis ends with four equations for the pitch curves, the first three of which are generally considered axiomatic. The fourth brings out an interesting relation between the radii from point of contact to center of rotation, the radii of curvature at the point of contact, and the angle between the pitch arcs and the line of centers at the point of contact, which might be of value in deriving a general method for constructing pitch curves for non-circular gears.

RELATIONS BETWEEN THE PITCH CURVES AND THE CURVES OF ACTION

The second article introduces the curves of action, or tooth profiles as they would now be called, and demonstrates their relation to the pitch curves. After proving the proposition, seldom neglected by modern textbook writers, that the normal to the acting curves passes through the pitch point,* two corollaries are stated, the first to the effect that the acting curves may be replaced by a line rolling on their evolutes, and the second a generalization of the notion of normal pitch. Both of these are well recognized in involute geometry; in fact, the first is a commonly-used principle for generating involute teeth, where the base circles are the evolutes of the tooth profiles. However, it is interesting to note their implications when generalized so as to apply to any form of gear tooth. It is also interesting that Gibbs, with characteristic thoroughness, explores the possibility of any exceptions to his propositions, and thus states that, when a pair of acting curves is replaced by another

* Gibbs is careful to distinguish between the line of action, defined as the common normal to the acting curves at their point of contact, and the locus of action, or curve described by the point of contact of the acting curves on the (fixed) reference plane, as M and N revolve. In general, the line of action would be a straight line passing through and rotating about the pitch point, while the locus of action would be a stationary curve. In involute gearing the two are of course identical.

pair ahead of, or behind, the first by a distance that is constant along the line of action, this distance must not include a center of curvature of the acting curves—e.g., the line of action in involute gearing cannot extend beyond the points of tangency with the base circles. It is apparent here as elsewhere that the author, however abstract his propositions may seem, never forgets that the acting curve is not a mere geometrical line but a true boundary between solid material on one side and empty space on the other.

Perhaps the most important part of this article is the establishment of two independent methods for constructing tooth profiles for a given pair of pitch curves. The first, which is elaborated in Sections 8 and 9, gives the centers and radii of curvature for the acting curves, for any chosen locus of action, by a simple geometrical construction. The author pays it scant attention, but it could possibly be developed into a useful means for drafting accurate profiles. The second is the time-honored method of rolling a curve (called "graphic curve" by Gibbs) on the pitch curves at the pitch point, and tracing the locus of a point (called "graphic point"), carried by the graphic curve, in the planes *M* and *N*. *Two* curves are thus generated simultaneously, and here, for the first time, the word "conjugate" comes into prominence.

The action of cycloidal teeth is always explained on this principle, because it is so obvious, but it is worth considering that other tooth forms may be generated in the same manner.* In fact, Gibbs gives the proposition underlying this method a most complete investigation (Sections 11 to 18) and proves,

* For example, if an equiangular spiral rolls on a straight line, its pole describes another straight line inclined to the first; hence, if an equiangular spiral be used as the graphic curve with two pitch circles, its pole will describe a straight line on the reference plane, and therefore two conjugate involutes on the planes *M* and *N*. However, this is clearly not a practical method for generating involutes, because it would require an infinite number of revolutions of the spiral to make its pole reach the pitch point. Gibbs recognizes the limitation by stating that the graphic-curve method becomes inapplicable as a *mechanical* solution whenever the acting curves cross the pitch curves at an oblique angle.

first, that *all* conjugate curves may be thus delineated, and second, that certain additional requirements must be met before conjugate curves thus produced may be classified as acting curves. For example, if the pitch curves are circles and the graphic curve a straight line, then the conjugate curves are involutes of the pitch circles and clearly inadmissible as acting curves except, as Gibbs points out, when one of the pitch circles becomes a rack, in which case a single point on the rack tooth will mesh correctly with the teeth of the opposite gear. The author also shows that, to generate curves by this method, the normals from all points of the curve must cut the pitch curve at least once.*

In conclusion, there is a brief remark concerning the "envelope" system of generating conjugate profiles, previously mentioned by Willis. It is only granted the status of an "approximation method," evidently with little thought that it was destined to become the basic principle in the design of all successful gear-cutting machinery in later years.

MODIFICATIONS OF FORCE INTRODUCED BY THE FRICTION OF THE ACTING SURFACES

This is a single brief section containing no analysis of relative sliding, but simply deriving an efficiency equation based upon a line of pressure which, due to friction, makes an arbitrary small angle with the line of action.

CASE OF CONSTANT RATIO OF VELOCITY

The pitch curves are now circles, convex towards each other, and tangent at *o* on the line of centers *mn*. Gibbs, with a characteristic tendency to generalize, assumes that the locus of action need not pass through *o* (though it is difficult to justify this after reading the argument of Section 15) but should do so in order to minimize friction. He then attacks the problem of describing the tooth profiles for a given locus of action, re-

* This had previously been noted by Airy. (Transactions of the Cambridge Philosophical Society, vol. 2, p. 278).

ferring first to a previous proposition on the location of their evolutes (Section 9), after which he demonstrates an interesting but somewhat involved rolling-curve method, whereby the graphic curve is first determined by means of its evolute, and then the acting curves are determined as roulettes of the graphic curve. Continuity of action is recognized by defining "stroke" as the distance traversed by a point on the pitch curve when the point of action moves from one end of the locus to the other, and a simple but completely general formula is derived for it.

INVOLUTE TEETH

This article contains the well-known facts that one would expect to find. In addition, two less well known matters are brought to light: (1) if tooth contact is restricted to the recess portion of the line of action, so that friction tends to draw the gears closer together, the efficiency may be improved to a theoretical limit of 100 per cent by increasing the pressure angle; (2) also by increasing the pressure angle, the "stroke" may be made as long as desired, so that a single-tooth pinion is possible. Fortunately, the author keeps his feet on the ground, dismissing the first as of academic interest only, since the pressure on the shaft bearings would be excessive, and limiting the second to motion in one direction.

EPICYCLOIDAL AND HYPOCYCLOIDAL TEETH

The author remarks upon the possibilities of varying the size of the graphic curve (circle in this case): for example, when it equals half the pitch circle, the acting curve is a radial line; when it equals the pitch circle, the acting curve is a point and its conjugate is an epicycloid. The lantern wheel and gear, common in clockwork, is the practical counterpart of this latter case, and Gibbs shows how the acting curves may be "mechanically described" by placing the tracing point on a link one end of which is pinned to the generating circle at what would otherwise be the graphic point, while the other end slides through the pitch point *o*.

A note on the equality of length of the "stroke" and the locus of action, in the case of this particular tooth form, concludes the article.

VELOCITY RATIO VARIABLE

This part of the thesis contains a very interesting development of a method for describing teeth such that the line of action shall always be normal to the line of centers. Gibbs discards the graphic-curve method, which though theoretically correct is impractical, and develops the evolute idea first discussed in Sections 8–10. The evolutes of the acting curves, which are nothing more than a generalization of the base circles in involute gearing, are denoted as "principal evolutes," and it is shown that if any "relations" (i.e., equations) are known for the radius, central angle, and radius-tangent angle for the pitch curves, these same relations apply to three specified dimensions of each principal evolute. The principal evolutes are therefore determined (either graphically or analytically as the reader wishes—the author does not trouble himself with such details) and the rest is simply a mechanical generation of the teeth by unwinding a tape, so to speak, from one evolute and winding it onto the other, as they and the pitch curves revolve. Gibbs forestalls the objection that this will not work when the tangent to the pitch curves is normal to the line of centers, by demonstrating a simple modification of the method whereby the line of action may be given any arbitrary inclination to the line of centers. This makes it possible to choose such an inclination that the angle between the line of action and common tangent to the pitch curves at the pitch point need never shrink to zero or become negative, even for a complete revolution of the gears. The author also gives a good example of the inverse construction; e.g., by taking two equal eccentric circles, or one large concentric and one small eccentric circle, as principal evolutes, teeth may be generated which give smooth action and a variable velocity ratio. It would seem that this inverse method has good practical possibilities, when the maximum and minimum angular velocity ratios are given and the precise distribution of intermediate ratios is unimportant.

TEETH OF CONSTANT PRESSURE; SUCCESSION OF TEETH; FRICTIONLESS TEETH; CIRCULAR ACTING CURVES

These concluding articles are, in the opinion of the writer, of little more than academic interest. The first leads to the conclusion that the pitch curves must consist of pairs of equiangular spirals, with one spiral of each pair set in reversed position with respect to the other; the second and third assume the possibility of discontinuities in tooth action which nowadays would be inadmissible, except perhaps in very special cases. The last one is a concession to the inaccurate workmanship of Gibbs' day—a concession that had been made by other theorists in this field, from Euler on down.

The present writer hesitates to pass any critical judgment on the value of the thesis as a whole to present-day gear designers or manufacturers, preferring to leave that to experts in this highly specialized branch of engineering. There is no doubt that much of it would be considered a review of well-established basic principles, and in many places the emphasis is placed on methods of tooth generation that have been tried out and found impractical. Following the lead of his predecessors, Gibbs sticks to the principle of generating the tooth profile by a tracing point instead of an enveloping line, and thereby fails to align himself with the march of progress that has brought the art of gear manufacture to its present high state of perfection. Of course, in his day the gulf between the academic theorist and the practical millwright had not been bridged; one may well wish that he might have had the stimulus of cooperation with modern gear engineers, competently trained and well versed in the geometrical problems of their industry, instead of having had to work alone. Nevertheless, the author's strict regard for logic, his ability to generalize a problem, and his devotion to the thorough study of a few fundamentals rather than their pursuit into all sorts of imaginable applications, mark his early efforts here as those of no ordinary intellect.

AN IMPROVED RAILWAY CAR BRAKE

AND NOTES ON OTHER MECHANICAL INVENTIONS

SAMUEL WILLIAM DUDLEY

In 1858, Willard Gibbs received his Bachelor's degree from Yale College and began his graduate work as a student of engineering in the Yale Scientific School which, three years later, was to be called the Sheffield Scientific School. His selection of applied science as a field of major study and of the scientific analysis of an optimum form for the teeth in spur gearing as the subject of his dissertation appear in retrospect to have been the natural responses of an alert and inventive mind to the exciting march of events within the college and in manufacturing and transportation.

After receiving his degree of Doctor of Philosophy in 1863, he was appointed Tutor in the College for three years, in Latin for the first two years and the third year in Natural Philosophy. His European sojourn, during which, for three years, he was to come under the profound influence of the leading mathematical physicists and chemists of the time, was still in the future.

Fragmentary evidence still extant shows that during his tutorship his mind was actively engaged in exploring both the theoretical and the practical aspects of a variety of subjects that attracted his attention. In all of these papers and memoranda, the penetrating and discriminating acuteness of his intellect was indicated. His notes include several pages of computations, sketches, and specifications relating to "a center vent" hydraulic turbine, accompanied by a mathematical analysis of the theory upon which were based the design and dimensions given. His early interest in basic concepts and fundamental scientific theory is evidenced by a "Discussion of a Set of

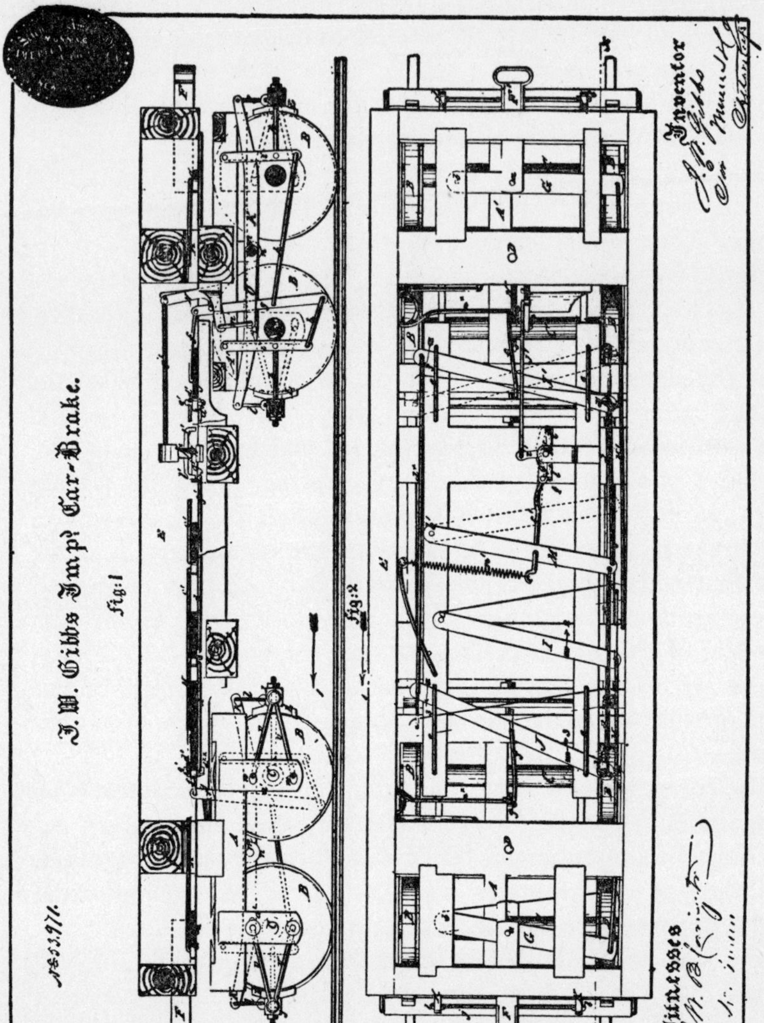
J. W. Gibbs Impd Car-Brake.
Fig:1
Fig:2
N°53,971.
Witnesses
Inventor

Figs. 1 and 2.

Units of Analytical Mechanics and Examples," incomplete and with some pages missing but showing throughout a clear grasp of essentials and an awakening interest in the analytical approach to fundamental problems which were to flower into transcendent philosophical genius under the stimulus of his European associations during the following three years.

The only original work which Gibbs seems to have completed while a tutor was a mechanical invention—an improved car brake.* The original letters patent for this device, handwritten, as issued by the U. S. Patent Office, were over the signatures of W. J. Otto, Secretary of the Interior, and T. C. Threaken, Commissioner of Patents. The accompanying specifications bear the signatures of the inventor, Josiah Willard Gibbs; the patent attorney, Munn and Company, New York City; and two witnesses, William F. McNamara and M. M. Livingston. Included is a top and a side view drawing, in ink on tracing cloth, witnessed by J. W. S. Covington and W. M. Trewin, of the brake mechanism as installed on the underframe and the two four-wheeled trucks of a railway car (see Figs. 1 and 2). These are bound together with a blue ribbon and sealed with the seal of the Patent Office, as was the custom of the day. There are no notes to indicate the date or the sources of this idea, or what led to his choice of this particular type of mechanism.

The correspondence with the patent attorneys includes: (1) a memorandum entitled "Nature and Object of the Brakes," explaining the purposes and functions which Gibbs had in mind. It also refers to "the model" which accompanied the application to the Patent Office and gives precise directions for its manipulation when unpacking, assembling, and demonstrating its operation, as required in the prosecution of the application through the Patent Office.

(2) A somewhat extended exchange of letters relating to the specifications and claims. There were five claims in a first draft

* U. S. Patent applied for October 30, 1865; Examined and allowed March 4, 1866; Granted April 17, 1866.

of the patent, of which only two were allowed, and two new claims were added in the patent as issued. On a preliminary memorandum of claims to be proposed, Mr. Gibbs writes:

"Special points of inquiry at the Patent Office

1. Does the construction of such brakes trespass upon any existing patent?
2. Can I claim all movable brakes? if not
3. Can I claim the use of this motion to break or change the connection of the coupling shaft with the brakes?"

The report from the Patent Office on the preliminary examination of his specifications and claims was dated "Washington, 30 October 1865." Mr. Gibbs was not satisfied with the brief reasons given by the Patent Office for the rejections, and he wrote his attorneys requesting copies of the two prior patent specifications cited by the Patent Office in support of their decisions. These copies were finally secured and are preserved with the patent as finally granted to Gibbs.

One was a rejected application from Edwin Amsden, Gainsville, New York, for a Buffer Brake for Truck (Railway Car), June 7, 1855, rejected June 30, 1855, with a prior art reference to a previously rejected application. This was from John Joseph Lahaye, Reading, Pennsylvania, for Railroad Car Brakes, February 15, 1854, rejected August 14, 1854. The rejection quoted a prior patent of Lahaye's own, dated April 10, 1847, reissued April 13, 1852, and also a still earlier rejected and withdrawn application from one Brooks, dated October 27, 1841, "and in many other rejections by this Office." These dates and prior art references are of special interest because they typify the activity of a large number of inventors in this field at the time.

All these brakes, including the Gibbs brake, belong to the class known as momentum brakes or buffer brakes. The fundamental purpose was "the dispensing with brakemen," so as to place the control of the train in the hands of the locomotive engineer. Actually, however, while the engineer could initiate an application of this type of brake by reducing the speed of the engine, thereafter he would have practically no control of the degree of braking on the cars, since the retarding force of the brakes on the cars depended upon the force with which the individual car pressed against the vehicle ahead, and this would vary according to the relative rate of retardation of adjoining vehicles, an extremely uncertain and uncontrollable factor.

Some form of train brake that would, in reality, be responsive to control from the engine was a desideratum eagerly sought by many inventors, both in the United States and abroad, from the early years of railway transportation, beginning with the mid-1830's and pursued with mounting vigor and imagination until the Vacuum Brake system of Naysmith and May (1844) and the Air Brake of Westinghouse (1869) provided the basic principles of practical systems capable of being adapted to the operating demands of railway transportation down to the present day.

Gibbs states in the memorandum,* "Nature and Object of the Brakes," already mentioned:

"These brakes are automatic. No brakeman is required for the cars nor has the engineer to pay any attention to them. He is supposed to have suitable means to check the velocity of the engine and tender and when he does this the automatic car brakes will keep the cars from pushing the locomotive onward by their momentum.

"The great advantage of this system over the hand-brakes (not to speak of the saving of the wages of the brakemen) is that the brakes may be put on the whole train instantaneously by simply checking the engine and with greater or less energy as the engineer may choose.

"Again cars of this system may be mixed up with others (ordinary

* The reader should bear in mind that this and the five succeeding paragraphs are not quoted from the patent specification but are notes and comments that appear to have been intended as preliminary information for the patent attorneys. Hence, they may not represent what would have appeared in a final draft after mature consideration. They are of interest, however, for their disclosure of the thoroughness with which Gibbs apprehended and analyzed existing railroad economic and operating conditions which his invention was intended to improve.

cars) in a train and the brakes will continue to act just as well as if all the cars had the automatic brakes. . . .

"It may be objected to this system that the brakes can never entirely counteract the momentum of the cars, so as to prevent them from pressing upon the engine, for, by the supposition, the engine must press back upon the buffer of the car, in order to put on the brakes. And it might be thought that in a long train the pressure of the cars upon the engine would be so great as to exert a considerable influence in keeping up the velocity of the latter. But it may easily be demonstrated that this action will not increase in proportion to the length of the train but will follow a law of the following nature.

"If the train consisted of but a single car behind the engine and tender, that car, supposed to have these brakes, would exert one-half the effect (to render the stopping of the engine more difficult) which a car without any brakes would exert. A train of two cars would have 3/4 of the effect which one car without brakes would have. A train of three cars would have 7/8 the effect of one car without brakes; a train of four cars 15/16, of five 31/32, etc., etc. So that no train, however long, of cars with these brakes can exert as great an effect to keep the locomotive from stopping as a single car without brakes.

"These figures would be somewhat altered by a different leverage in the mechanism of the brakes, but the general result would be the same."

The reasoning in these last two paragraphs is not easily followed, there being no indication of the assumptions upon which it is based. Mr. Gibbs recognizes that "the leverage in the mechanism of the brakes" will have some effect, but, whatever the leverage, "the engine must press back upon the buffer of the car" in order to produce any braking effect at all upon the car. As soon as the braking causes the car to slow down, the pressure between the engine and the car would be reduced. This would, in turn, cause the braking effect on the car to decrease, since its source of "pressure" has become less, and the car then tends to decrease its speed less rapidly, causing the engine to press back more heavily against the car, thus tending to increase the braking effect upon the car, and so on. In this manner alternating buffing and pulling shocks might be set up between the cars (especially with the link-and-pin type of drawbar connection of that day), producing conditions that could hardly be expected to follow any such simple "law" as suggested. If the ratio series given above is continued for a train of ten cars, the conclusion would be that such a train would exert 1023/1024 of the effect upon the engine which a car without any brakes would exert.

Since the brake action will be influenced by the leverage employed, by the arrangement of "the oblique rods, *v* (Fig. 1)," by the coefficient of friction between the brake shoes and the wheels and especially by the variable reactions between the moving vehicles, the resultant retardation of the car might be expected to vary widely. Alternate buffing and pulling actions while stopping might well result in rather sudden variations in the retardation of the car from an amount greater than that on the engine to almost nothing at all.

One is led to wonder if speculation upon the possible behavior of several loosely connected cars, having considerable free slack as well as spring resistance between them, might not have been the origin of another idea, disclosed by an undated memorandum (probably *c.* 1866) and sketch found among the Gibbs papers, which reads as follows:

"COUPLINGS"

"Friction may advantageously be used to augment the effect of springs (coupling) in bearing the force of the jar of slight collision."

"Thus"

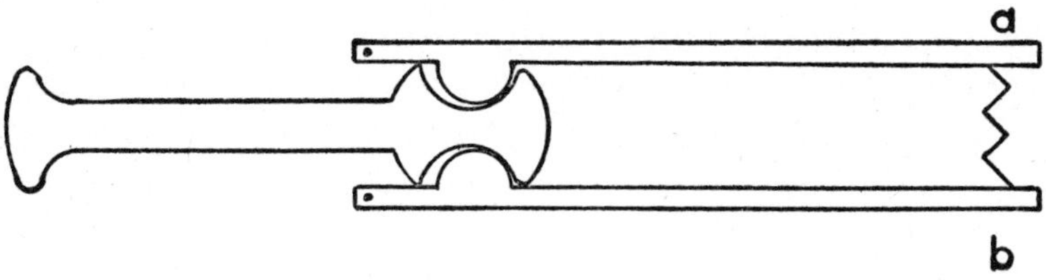

Fig. 3. Improved Railway Car Brake.

"The ends of the levers at *a*, *b*, are to be drawn together by powerful springs."

This simple construction comprises the essential working elements of a friction draft gear. It appears to have been an entirely original invention of a frictional (energy absorbing) device, to replace the usual spring (energy storing and restoring) means for mitigating the annoying and often damaging jerks and shocks between cars, inevitable in trains of more or less loosely coupled units. George Westinghouse suggested, a dozen years later, the advantages of dissipating the energy of shocks between cars through work against frictional resistance instead of cushioning by a spring which, after being compressed, would give back a substantial portion of its stored energy. It was not until 1888, however, that a practicable design of Friction Draft Gear was developed, and now some form of this device is standard equipment on American railroads.

This fragmentary evidence of Mr. Gibbs' clear vision, at so early a date, of a fundamental need in railway train operation is particularly interesting in view of the simplicity and effectiveness of his solution by the utilization of frictional resistance to accomplish the desired result.

Returning now to the Gibbs brake system, Fig. 1, the following is a summary of the description of its operation, as given in the patent, the quotes indicating wording taken verbatim from the patent specification: "This invention has for its object the dispensing with brakemen and to arrange the working parts of the brake in such a manner that the brakes of all the cars comprising a train will be simultaneously applied by the momentum of the train whenever the speed of the locomotive is checked, and the brakes thrown off whenever the locomotive has its speed increased or is started from rest, the above results being attained whether a train be moving forward or backward.

"'When the speed of the locomotive is checked the momentum of the cars will press or push the drawhead inward'—then, through a connected system of levers and rods, the brake shoes 'will be pressed against the wheels.'

"Thus it will be seen that by this simple arrangement a perfectly automatic brake is obtained, so far as the forward movement of the train is concerned, but it will also be seen that without some suitable appliance the train cannot be backed, as the shoes will press against the wheels."

The patent then describes an ingenious arrangement of rods, levers, shafts, a "three armed sleeve, h^x," and springs, whereby the elevation or depression of the shoes, locked against the wheels by the pushing in or pulling out of the drawheads, will act automatically to change-over and lock the rigging, for applying and releasing the brake in the position required to allow the brakes to function in the manner intended, whether the train is being pulled in forward motion or pushed in backing motion.

Three of the four claims allowed in the patent cover the combination and arrangement of the structural members required to accomplish these automatic change-over and locking functions. This automatic change-over functions properly, the patent points out, only so long as the locomotive is at the end of the car or train assumed thus far (indicated by the arrow, Fig. 1). If the locomotive is at the other end of the car, the lever system must be reset to a new configuration by manually disconnecting the lever *M* from the operating rod *H*, previously in use, and attaching this lever to the corresponding operating rod *H*, extending to the end of the car where the locomotive is now located.

The fourth claim relates to "oblique rods, *v*," attached to the "shoe bars, *j*" (the brake beams). These rods, *v*, limit the permissible vertical motion of the shoes and "on account of their obliquity, very much increase the pressure of the shoes upon the wheels."

There remain no records to indicate that Gibbs gave any further consideration to his brake after having received his patent. Later in the year 1866 he was on his way to France and Germany, where the new vistas and wider horizons of mathematics and the physical sciences awaited him. Had his work on the railway car brake problem foreshadowed to any degree the quality of those perceptive, analytical, and creative powers that so notably distinguished his later work? Considering the limited opportunities presented to him during his tutorship and the state of the art of railway train control at the time, the wonder is not that his invention did not markedly affect the practices of railway inventors and engineers but that it did exhibit a recognition of so many of the essential requirements of a brake system that would provide for the safe, economic, and efficient control of a moving train.

The fundamental requirements which Gibbs sought to satisfy, and did to a degree provide for, are by no means insignificant nor are they any the less essential, although they are more generally recognized today than in his time. Gibbs refers to many of these specifically; for example:

(1) The brakes must enable the engineer to control the speed of the train at any time, to any desired degree.

(2) They must operate with equal effectiveness in either forward or backward motion of the train.

(3) They must function equally well with the locomotive at either end of the train.

(4) They must be automatic in action (by which Gibbs meant that, once properly adjusted, they should respond automatically to the retardation of the locomotive created by the engineer's manipulation of the brakes on the locomotive, without requiring any action by the brakeman on the car).

(5) They must be simultaneous in operation on all cars in the train—a requirement far from being realized, of course, by Gibbs' brake, nor can it be completely satisfied by any system of brakes (except when electrical control is used), although modern pneumatically controlled train brakes approximate this requirement sufficiently for practical railroad operation.

(6) They must function when mixed in trains with cars not equipped with these brakes—a particularly significant requirement, as noted by Gibbs, and one that cannot be met by the modern types of train brakes as is evident from the Interstate Commerce Commission ruling that trains may not leave their

terminals until 100 per cent of the cars composing the train are known to be equipped with operative brakes. It must be noted before leaving this point, however, that although the buffer brake of Gibbs, or any other buffer brake, would operate on its own vehicle when mixed in a train with non-equipped cars, the result to the train as a whole might well be more disastrous than no brake at all.

(7) They should improve the braking effect on the train in proportion to the number of vehicles so equipped—a point to be noted here because, while it may seem obvious, it illustrates how systematically and thoroughly Gibbs had analyzed the elements composing his problem.

(8) The energy of the shocks between adjoining and more or less loosely coupled cars should not be cushioned by spring compression, to be given back in considerable and sometimes damaging amounts, to the structures upon recoil, but should be allowed to do work against frictional resistance, thereby absorbing and dissipating the impact energy and eliminating recoil.

This last fundamental desideratum was an original, direct, simple and scientific solution of a serious difficulty with which railroads struggled until the friction draft gear was developed a quarter of a century later.

The means employed by Gibbs to satisfy the requirements for satisfactory train control were similar, in many respects, to those adopted by other inventors, as shown by the prior art quoted on his disallowed claims, but the features covered by the claims that were allowed show a high degree of ingenuity and demonstrate definitely a thorough study of the provisions necessary to insure that the mechanism would meet many of the requirements of practical operation.

That so many of these items were apprehended and, to a degree, provided for, shows an insight into and grasp of the basic principles and the practical requirements of railroad train control which would hardly be expected from a tutor of Latin and Natural Philosophy in Yale College, who afterward became the greatest mathematical physicist of his age.

THE GIBBS GOVERNOR FOR STEAM ENGINES

LYNDE PHELPS WHEELER

In order to appreciate the originality and ingenuity of this invention it is necessary first to envision the conditions which must be fulfilled by a device designed to hold automatically the rotational speed of a prime mover such as a reciprocating steam engine to a constant predetermined value, and second to consider how far these conditions are met by the Gibbs invention as compared with those of other inventors.

The general principle of operation of all of these devices is to produce an alteration in the power supplied to the engine which will compensate for unavoidable fluctuations in the power output. Thus, in the case of the steam engine, if for any reason the load into which the engine works is increased or if the boiler pressure falls—one or both—the engine will tend to slow down and it is the function of the governor to increase the steam supply until the normal running speed is restored. Conversely, in the case of a decrease in external load and/or a rise in boiler pressure, the governor must act to reduce the steam supply. Since up to recent times mechanical devices were the only ones employed to implement this general principle, it will not be necessary to consider here any other class of device.

Now the mechanical devices which have been used to perform the governing function all operate through the force which it is necessary to supply in order to control the centrifugal tendency of masses rotated by the engine. These masses (commonly called the governor "balls") are usually two in number in order to avoid unbalanced stresses on the rotating spindle which carries them. The required controlling force may be supplied by the weight of the masses alone, by springs alone or by a combination of the two.

In general there are only two types of these devices—first, those utilizing the principle of the conical pendulum in which the vertical component of the displacement of the balls is caused to operate on the power supply; and, second, those in which the motion of the masses is confined sensibly to one plane and the radial displacement of the masses in that plane controls the power supply. In governors of the first type—to which the Gibbs invention belongs—the controlling force employed has usually been that due to the weight of the balls alone, although in some forms this has been supplemented by spring control. In governors of the second type the controlling force is supplied solely by springs of adjustable tension.

In either type the magnitude of the controlling is determined from the fundamental law of motion in a circle

$$F = 4\pi^2 N^2 RM, \tag{1}$$

where M is the mass and R the radius of the circle in which it moves at a speed of N revolutions in unit time.

Hence if any small fluctuation in N occurs, say δN, the percentage fluctuations in F and R corresponding to those in N are connected by the relation

$$\frac{\delta F}{F} - \frac{\delta R}{R} = 2\,\frac{\delta N}{N}\cdot \tag{2}$$

From this it follows that in order to keep $\delta N/N$ small, the percentage change in the controlling force should exceed that in the radius by as small an amount as possible. If $\delta F/F$ should be made exactly equal to $\delta R/R$, no change in N could take place at any finite running speed. In other words there would be but one speed and radius at which the revolving mass could be in equilibrium with the controlling force and any departure from that speed would cause an indefinitely great change in R. A governor in which the condition of equality in the rates of increase of controlling force and radius is approximated closely is said to be *sensitive* and to approach *isochronism*. The margin

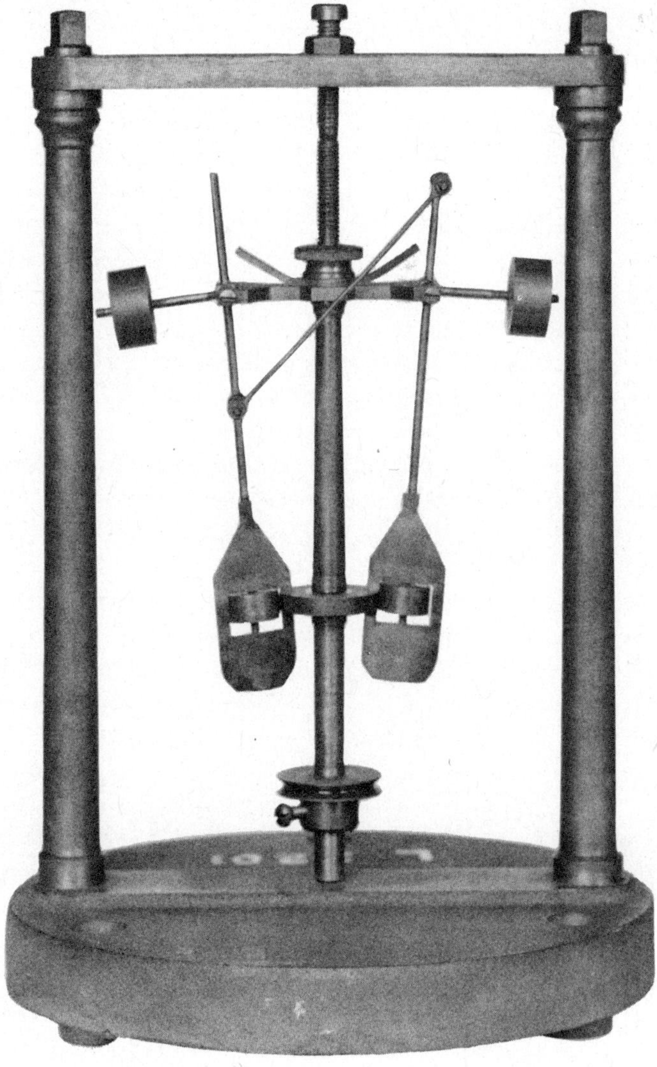

MODEL OF THE GIBBS GOVERNOR
On permanent display in the Sloane Physics Laboratory

of excess of $\delta F/F$ over $\delta R/R$ is taken as a measure of the *stability* of the governor.

Now while isochronism would be an ideal characteristic in a governor, the accompanying violent changes in radius which would then occur in response to the slightest change in speed would be undesirable in practice. Hence the object of the designer is to obtain as close an approach to isochronism as is compatible with a margin of stability sufficient to secure that this condition of forced oscillation which is known as "hunting" is kept within reasonable bounds. (Some hunting, due to the inertia of the moving masses, is of course inevitable.) That this objective can be quite simply and practically attained in those governors where all or a portion of the controlling force is supplied by springs of adjustable tension is sufficiently obvious, and has been the method most commonly used, especially for the higher speed prime movers.

That the same end can be attained without resorting to spring control is not so immediately obvious. To effect the adjustment of the controlling force without such aid has posed a problem which has evoked much ingenuity on the part of many inventors. All of the solutions of this problem (including that of Gibbs) have consisted in modifications of the original conical pendulum of Watt designed to secure greater sensitivity. To obtain the necessary margin of stability it has been the usual practice to employ auxiliary devices coming into play when the amplitude of oscillation of the balls exceeds predetermined limits. It is one of the merits of Gibbs' invention that such extraneous aids are not theoretically required; adjustments of the pendulum itself furnishing a simple method of control *both* of sensitivity and margin of stability.

In order to make these developments intelligible and to place the Gibbs governor in an adequate perspective it is necessary first to consider the theory of the simple conical pendulum governor and then that of two of the most important ways in which greater sensitivity has been obtained without the use of spring control. Since the Gibbs governor was never actually applied to an engine and exists only as a model illustrating the bare essentials of the principle involved, we will compare it only with simplified or idealized forms of the other devices considered.

The oldest of these is that of James Watt. It consists of two masses carried by rigid arms hinged to a vertical spindle which is rotated by the engine. Schematically it may be represented as in Fig. 1, the hinges being at the points P. Qualitatively the operation is simple to understand. Starting from rest (zero angular speed of the spindle), when the balls may be supposed

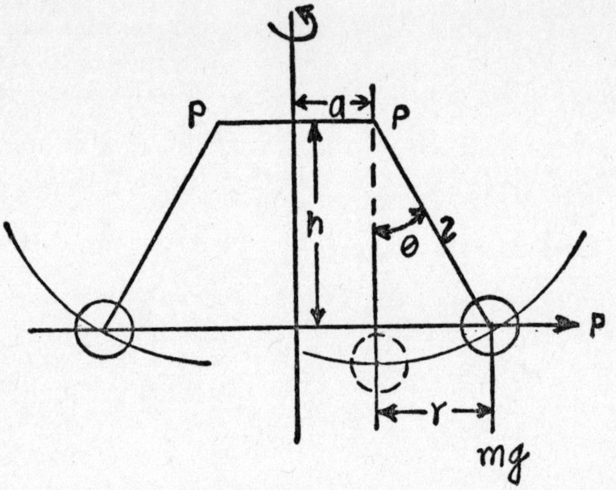

Fig. 1. Schematic Diagram of the Simple Watt Governor.

ideally to hang vertically ($\theta = 0$), as the speed is increased the rotating balls move outward and upward along circular arcs in a vertical plane, each value of the radius r passed through corresponding to a definite rotational speed. Thus within limits determined by the geometry of the figure and constructional considerations dependent on the particular method employed to transmit the vertical rise of the balls to the throttle or the valve mechanism, this form of governor will run stably at any given speed and changes in speed will produce linear displacements which can be made to make compensatory changes in the power input. It is not necessary for the present purposes to consider further the mechanical details of the mechanism for actuating the throttle or valve travel as this varies widely in practice and does not affect the fundamental mode of operation of the pendulum.

Quantitatively the action may be described from the equilibrium which must subsist between the torque due to the centrifugal force F (equal in magnitude and opposite in direction to the controlling force) and that due to the weight of a ball, mg. Thus, from the equality of the torques, the fundamental law of circular motion (equation 1) and the geometry of the figure, we have

$$Fh = mgr, \quad \text{with} \quad F = 4\pi^2 N^2 (r + a)m, \quad h = l \cos\theta, \quad r = l \sin\theta.$$

From these relations it follows that the equilibrium speed is

$$N = \frac{1}{2\pi}\sqrt{\frac{g}{h}} \cdot \sqrt{\frac{\sin\theta}{\sin\theta + a/l}} = \frac{1}{2\pi}\sqrt{\frac{g}{l}} \cdot \sqrt{\frac{\operatorname{tg}\theta}{\sin\theta + a/l}}\,.$$

For the present purpose of comparison with modified forms of the governor it is convenient to replace N by what may be called the "specific" speed, that is, one independent (to the first order in the small quantity a/l) of the actual dimensions of the governor. This we will define by

$$n = 2\pi N\sqrt{\frac{l}{g}},$$

making the equation for the equilibrium speed read

$$n = \sqrt{\frac{\operatorname{tg}\theta}{\sin\theta + a/l}}\,. \tag{3}$$

A plot of this equation computed for $a/l = 1/3$ is shown in curve 1 of Fig. 4. This curve (or the equation) shows clearly the essentially stable nature of the device; that is, that to every value of the radius there corresponds a different equilibrium speed. It also permits an estimate of the feasible range of speeds (for any given value of l) for which it is suitable. With dimensions such as are customarily used and remembering that constructional considerations limit the applicability of the curve to the central portion on either side of $\theta = 45°$, it can be computed that the device is practicable only in the relatively low speed range of from 50 to 70 revolutions per minute. This speed may be increased (and has been in practice) some two or three fold by the device of "loading," that is by adding a mass sliding on the spindle which exerts no torque about the points P, thus effectively increasing the weight of the balls without increasing their mass. However, since the same device can be applied to any form of conical pendulum governor, it is not necessary to consider it further here where we are only concerned with the *relative* performance of the different forms.

Next, in order to evaluate the sensitivity of this simple form of governor, we define the "specific" sensitivity as the percentage change in the height h per unit change in the specific speed, or

$$S = \frac{1}{h}\frac{\delta h}{\delta n}, \tag{4}$$

which by equation (3) and the value of h in terms of θ yields

$$S = -\frac{2}{n}\cdot\frac{\sin\theta + a/l}{\sin\theta + \dfrac{a}{l}\csc^2\theta}. \tag{5}$$

A plot of this equation for the particular value of $a/l = 1/3$ is shown in curve 1 of Fig. 5. This curve (or the equation) shows first that an increase in θ (that is an increase in speed) always means a decrease of h. Second it shows that the sensitivity has a maximum at an angle of about 51° measured from the ideal rest position. This angle then fixes the position which should be made to correspond to the normal running speed of the engine. Finally it follows that the sensitivity is inversely proportional to the speed and cannot be materially increased by any adjustment of the ratio of a to l or other change in dimensions. With dimensions such as are ordinarily met with in practice it can be deduced from the equation that the maximum sensitivity is such that a change of one revolution per minute will cause a change in h of about one quarter of an inch.

Thus equations (3) and (5) teach us that the simple Watt governor is an essentially low speed, stable, and relatively insensitive device, conclusions which are amply borne out in practice.

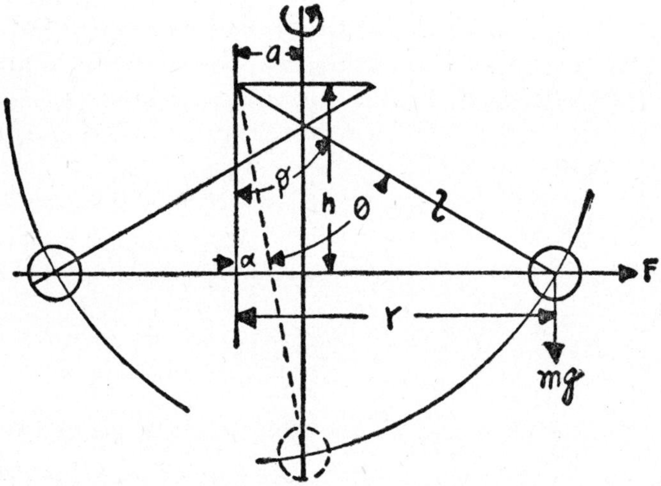

Fig. 2. Schematic Diagram of the Crossed Arm Governor.

The search for means of improving the sensitivity of the conical pendulum have for the most part been along the lines of constraining the balls to move along parabolic rather than circular arcs in the vertical plane. If this is done, the invariable subnormal of a parabolic arc assures that there will be but one speed at which such a governor will run stably and that the sensitivity will in consequence be indefinitely great.

While ingenious arrangements of cams and links have been devised to accomplish this result with precision, the method which has been used most frequently has been to make the naturally circular vertical arcs of the balls approximate that of a parabola over a short range of arc. This can be done by placing the centers of suspension of the ball arms as closely as possible to the instantaneous center of a parabolic arc coinciding, at the position corresponding to the normal running speed, with that of the actual circular arc. This results in some one of the many forms of the so-called "crossed arm" governor, which is shown schematically in Fig. 2.

Here the rest position of the balls (zero angular speed of the spindle) may be taken (ideally) with the ball arms making an angle α with the vertical whose sine is given by a/l. From the condition of torque equilibrium and the definitions of n and S as before, we deduce

$$n = \sqrt{\frac{\operatorname{tg} \vartheta}{\sin \vartheta - a/l}} \tag{6}$$

and

$$S = -\frac{2}{n} \cdot \frac{\sin \vartheta - a/l}{\sin \vartheta - \dfrac{a}{l} \csc^2 \vartheta} . \tag{7}$$

Plots of n vs θ and S vs θ ($\theta = \vartheta - \alpha$) with $a/l = 1/3$ are shown in the curves 2 of Figs. 4 and 5. From the first of these we see that starting from the ideal rest position the balls move from a region of unstable equilibrium (n decreasing with increasing θ) through a short region—around $\theta = 23.5°$ in this particular case—through which the speed does not change appreciably, and finally into a region of stable equilibrium, where every value of θ corresponds to a different speed. The second curve (curve 2 of Fig. 5) shows the indefinitely great increase in sensitivity that occurs in passing from the unstable to the stable region, that is in the region of approximate isochronism.

If the value of a is altered the region of approximate isochronism will be shifted to a new angular position and over the previous range will show a different speed for every value of θ. That is, for a given normal running speed changing a converts the governor to the stable insensitive type and such an adjustment might be used to provide the required margin of stability. However, constructional considerations render an adjustment of this kind difficult to provide in practice, so that in general the control of hunting in this type of governor is accomplished by frictional devices brought into play at given speed limits. It should be noted that such devices (dash pots, for instance) represent what may be termed afterthoughts, introduced to make the improvement in the sensitivity of the simple governor operative in practice.

Other solutions of the problem of increasing the sensitivity of the conical pendulum governor have followed the line of providing means of directly altering the relative rates of increase of controlling force and radius (see equation 2), instead of constraining the balls to move in more or less approximate parabolic arcs. It will only be necessary here to consider briefly the most ingenious of these, that of Dr. Pröll which, incidentally, came out in 1872 and is therefore practically contemporaneous with the Gibbs invention. Pröll's governor is an inverted conical pendulum with the constraints which are necessarily introduced by the inverted structure so arranged as to provide a passive torque which, over a portion of the arc of operation, opposes that of the balls, thus reducing the overall controlling force. It is intrinsically a governor of the "loaded" type. The quantitative theory of the device can be developed similarly to the previous cases, but as the resulting curves of n vs θ and S vs θ resemble very closely those shown in Figs. 4 and 5 for the crossed arm governor, it does not seem worth while to go into it in detail here. The principal differences revealed by such an analysis are a shorter region of initial instability and a longer region of approximate isochronism. While it is theoretically possible with this governor to include adjustments for the position of the balls which could be used to control the degree of approximation to isochronism and thus to provide a suitable margin of stability, these are not so simple as those possible with the Gibbs governor and in practice the control of hunting is usually secured by accessory devices as in the case of the crossed arm governor.

Gibbs' approach to this problem of improving the sensitivity of the simple conical pendulum was quite different from those just discussed. Instead of regulating the controlling force by constraining the balls to move in more or less approximate vertical parabolic arcs or by utilizing passive torques due to constructional constraints, he conceived the idea of proportioning the increase of controlling force to that of the radius by providing

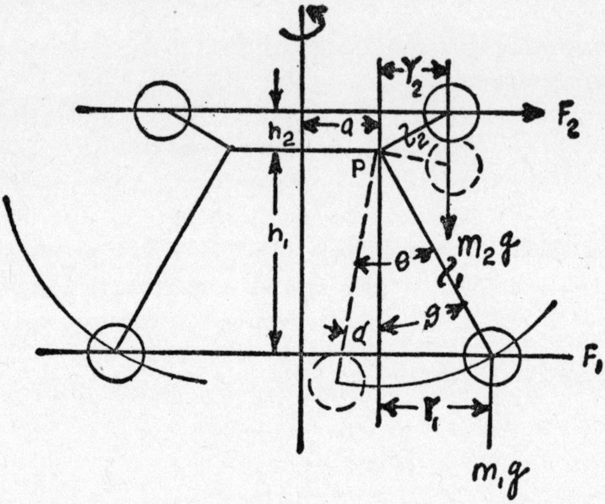

Fig. 3. Schematic Diagram of the Gibbs Governor.

a second *active* centrifugal torque which for a portion of the vertical arc would oppose that of the balls of a simple Watt governor. As can be seen from the photograph of the model facing page 64, or as shown schematically in Fig. 3, this is accomplished by providing a second pair of balls. The balls of each balancing pair are mounted at the extremities of a right angled bell crank lever which is hinged at the junction of the supporting arms to a cross piece carried by the rotating spindle. In the (ideal) position of rest—zero angular speed of the spindle—the bell crank takes up a position such as that indicated by the dotted lines in Fig. 3, defined by an angle whose tangent is given by the ratio m_2l_2/m_1l_1. Starting from this rest position the torques about the point P due to centrifugal action have the same sign for each ball of the pair. But as the speed increases until the lower arm of the bell crank lever passes the vertical, the two torques become of opposite sign, thus producing a decrease in the controlling force which can be adjusted to any desired value (within limits) by altering the positions of the balls on the bell crank arms whose ends are threaded for this purpose.

Quantitatively the equilibrium of the torques acting about P yields the equation

$$F_1h_1 - F_2h_2 = g(m_1r_1 + m_2r_2),$$

with,

$$h_1 = l_1 \cos \vartheta, \quad h_2 = l_2 \sin \vartheta, \qquad r_1 = l_1 \sin \vartheta, \quad r_2 = l_2 \cos \vartheta,$$

$$F_1 = 4\pi^2N^2(r_1 + a)m_1, \qquad F_2 = 4\pi^2N^2(r_2 + a)m_2.$$

From these relations we may write the equation for the equilibrium speed in the form

$$N = \frac{1}{2\pi}\sqrt{\frac{g}{L}\cdot\frac{A}{B}}$$

where,

$$L = l_1 - l_2 \operatorname{tg} \alpha, \qquad A = \sin \vartheta + \operatorname{tg} \alpha \cos \vartheta,$$

$$B = \sin \vartheta \cos \vartheta + \frac{a}{L}(\cos \vartheta - \operatorname{tg} \alpha \sin \vartheta). \tag{8}$$

If as before we introduce the notion of "specific" speed, defined now by

$$n = 2\pi N\sqrt{L/g},$$

the equation becomes

$$n = \sqrt{A/B}, \tag{9}$$

from which and the definition of the "specific" sensitivity (equation 4) we may write

$$S = \frac{C}{AB} \tag{10}$$

where,

$$C = \sin^3 \vartheta - \text{tg}\,\alpha \cos^3 \vartheta + \frac{a}{L} \sec^2 \alpha. \tag{11}$$

Now from measurements made on the model it appears that Gibbs intended that $m_2 l_2$ should be equal to $m_1 l_1$, or $\text{tg}\alpha = 1$, with $m_2/m_1 = 3$. From the range of adjustment provided it is also evident that this intention can be fulfilled and at the same time l_2 can be made as nearly equal to a as may be desired. With these values suggested by the model equations 8 and 11 read

$$\begin{aligned} L &= l_1 - l_2, \qquad A = \sin \vartheta + \cos \vartheta, \\ B &= \sin \vartheta \cos \vartheta + \tfrac{1}{2}(\cos \vartheta - \sin \vartheta), \\ C &= (\sin \vartheta - \cos \vartheta)(1 + \sin \vartheta \cos \vartheta) + 1. \end{aligned} \tag{12}$$

Plots of n vs θ and S vs θ from equations 9 and 10 ($\theta = \vartheta + \alpha$), with the values of L, A, B and C as given in equation 12, are shown in curves 3 of Figs. 4 and 5. From the first of these it is seen that starting from the ideal rest position ($\theta = \alpha$) the motion of the balls in the vertical plane is at first one characterized by stability (a discrete speed for each θ), followed by a relatively long arc of very approximate isochronism (little change in speed with θ) and finally by a second region of stability. From the second curve (curve 3 of Fig. 5) we see the same indefinitely great increase in sensitivity at a particular angular position (45° in this case) as is exhibited by the crossed arm governor. The most marked difference between the two types of governor as revealed by the curves is the greater extent of the region of approximate isochronism shown by the Gibbs governor, and it is because of its superiority in this respect that it is possible to secure a higher sensitivity with the necessary margin of stability.

This is illustrated by the curve 3a of Fig. 5, which is drawn for a value of $l_2 = 0.99a$ instead of the value $l_2 = a$ for which curve 3 is drawn. This value of l_2/a is readily obtainable with the range of adjustment provided for in the model while still re-

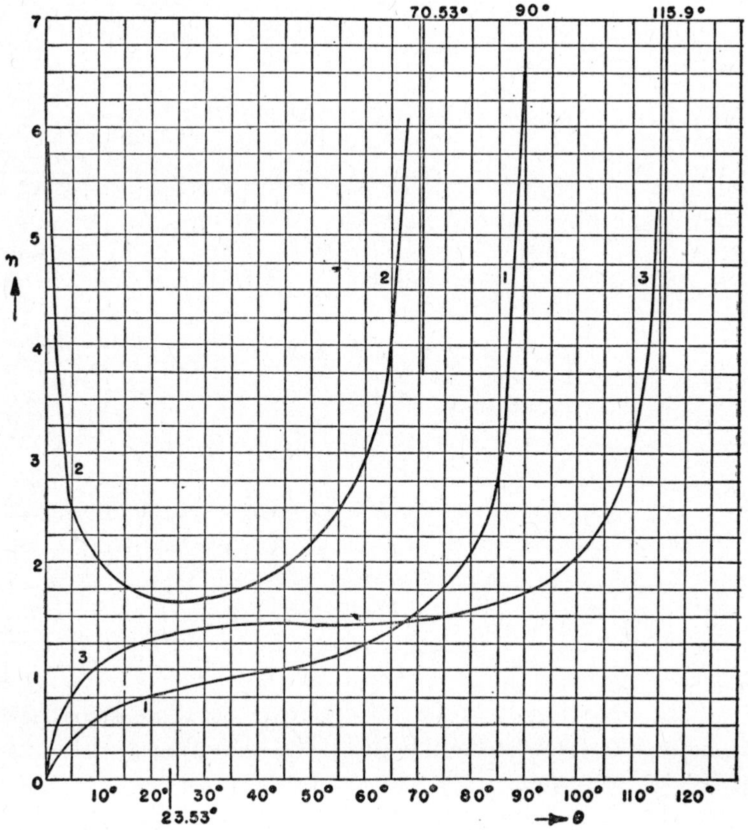

Fig. 4. Equilibrium Speed vs. Angular Position (measured from the ideal rest position). *Curve 1:* The Simple Watt Governor. *Curve 2:* The Crossed Arm Governor. *Curve 3:* The Gibbs Governor.

taining the same value of l_1/l_2. A form of sensitivity curve like 3*a* showing a continuous variation through a large *finite* sensitivity does not exist for any other form of conical pendulum governor. All the others show a discontinuous change through an *infinite* value at the position of maximum sensitivity. It is this unique feature of the Gibbs governor that permits the simultaneous adjustment of both the sensitivity and stability by simple adjustments of the position of the balls. It is true that the model shows that some small damping is provided for in the air vanes associated with the lower balls. These however are not essential to the attainment of the requisite margin of stability (as has been shown above), and merely permit a slightly higher sensitivity than would be practicable in their absence.

In one other respect the Gibbs governor has possibilities transcending those of any previous conical pendulum device. This is because it is intrinsically a higher speed device owing to the fact that its equilibrium speed is inversely proportional to the root of the difference in the lengths of the arms of the bell crank rather than to that of the whole length of a single arm as is the case with other forms of this type of governor. Thus if the l of Fig. 1 were the same as the l_1 of Fig. 3, then with the Gibbs governor adjusted as apparently intended in the model ($l_1/l_2=3$), the equilibrium speeds n would be some 25 per cent greater than those attainable with the Watt governor. With suitable changes in the proportioning of the masses and the lengths of the bell crank arms this advantage could be very considerably increased.

In view of the obvious advantages of the Gibbs invention it may seem to some to be unfortunate that the idea has never been exploited in practice. The reasons why Gibbs himself never pushed his invention even to the extent of publishing a description of it or seeking a patent on it are not however far to seek. In the first place it would have taken considerable capital and time-consuming effort to interest engineers and manufacturers in a device which after all would only do somewhat better something that already was being done in a sufficiently satisfactory manner in most instances. Novelty and ingenuity are not the only or even the most important factors determining the successful exploitation of any invention. The "reduction to practice" demands an incentive and interest which at the most were but feebly operative in Gibbs' case.

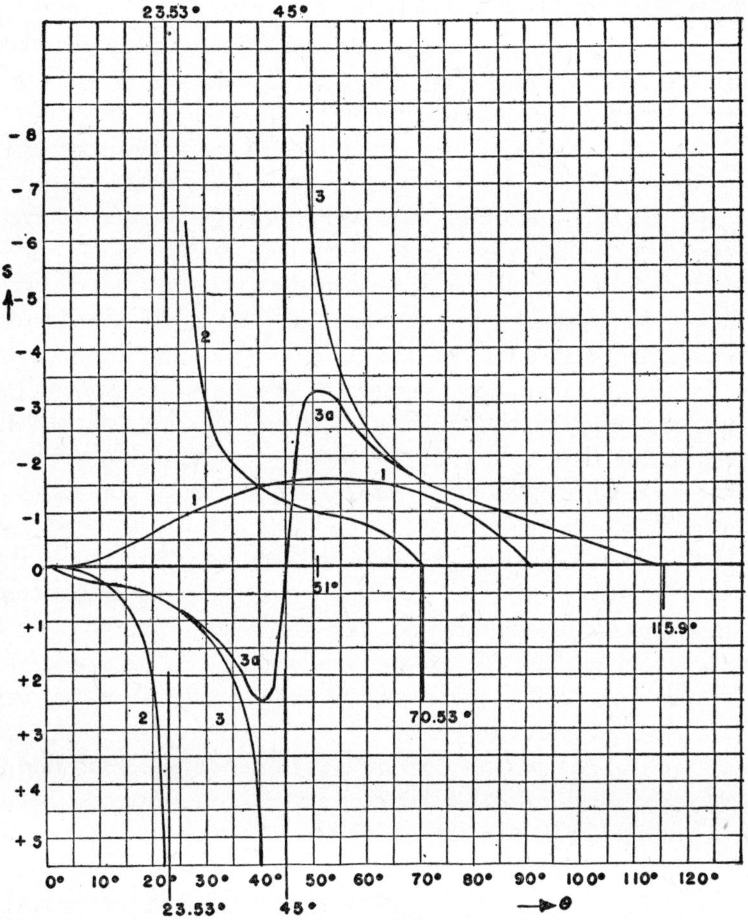

Fig. 5. Sensitivity vs. Angular Position (measured from the ideal rest position). *Curve 1:* The Simple Watt Governor. *Curve 2:* The Crossed Arm Governor. *Curve 3:* The Gibbs Governor with $a/l_2=1$. *Curve 3a:* The Gibbs Governor with $a/l_2=0.99$.

Moreover, this idea came to him at a time (as pointed out in the Introduction) when he was beginning to lose his earlier interest in applied mechanics and to become more and more absorbed in the more profound mysteries of molecular mechanics to whose elucidation he was destined to make such fundamental contributions. Thus it is not surprising that he made so little effort to interest others in his invention.

It was not with the idea of inducing anybody at this late date to undertake the practical development of the Gibbs governor that the present analysis was made. With the advent of the electric dynamo, the internal combustion engine and the turbine (first steam and now gas) and the resultant much higher speeds of revolution than were contemplated at the time when Gibbs' invention was made, and for the regulation of which the conical pendulum principle is not suited, there is not much of practical interest or importance in Gibbs' idea.

But if the practical importance of this work of the late sixties or early seventies of the last century scarcely justifies its publication in 1947, the intrinsic interest of the method of thought it exhibits as the earliest example of Gibbs' use of that powerful process of generalization which underlies the enduring quality of his great contributions to our understanding of the properties of matter, seems to me to render its publication important and appropriate. Here in this work on the governor he took the well-known principle of the equilibrium of two torques and generalized it by adding a third. This is precisely the same process followed in the monumental work on the Equilibrium of Heterogeneous Substances, where he took the well-known principle of the equilibrium between the various forms of energy subsisting in a homogeneous substance and generalized it by adding terms of the same form to determine the energy equilibrium in a heterogeneous mass: and thus was born the science of Physical Chemistry. The importance of this and other later applications of this method of generalization of course completely overshadows that of the early work we have been considering, but it is interesting and significant that some of the techniques of Gibbs' immortal canvases are discernible in this minor and almost forgotten example of his youthful interests.